T0137081

Studies in Systems, Decision and Control

Volume 201

Series editor

Janusz Kacprzyk, Systems Research Institute, Polish Academy of Sciences, Warsaw, Poland
e-mail: kacprzyk@ibspan.waw.pl

The series "Studies in Systems, Decision and Control" (SSDC) covers both new developments and advances, as well as the state of the art, in the various areas of broadly perceived systems, decision making and control–quickly, up to date and with a high quality. The intent is to cover the theory, applications, and perspectives on the state of the art and future developments relevant to systems, decision making, control, complex processes and related areas, as embedded in the fields of engineering, computer science, physics, economics, social and life sciences, as well as the paradigms and methodologies behind them. The series contains monographs, textbooks, lecture notes and edited volumes in systems, decision making and control spanning the areas of Cyber-Physical Systems, Autonomous Systems, Sensor Networks, Control Systems, Energy Systems, Automotive Systems, Biological Systems, Vehicular Networking and Connected Vehicles, Aerospace Systems, Automation, Manufacturing, Smart Grids, Nonlinear Systems, Power Systems, Robotics, Social Systems, Economic Systems and other. Of particular value to both the contributors and the readership are the short publication timeframe and the world-wide distribution and exposure which enable both a wide and rapid dissemination of research output.

More information about this series at http://www.springer.com/series/13304

Ralf Stetter

Fault-Tolerant Design and Control of Automated Vehicles and Processes

Insights for the Synthesis of Intelligent Systems

 Springer

Ralf Stetter
University of Applied Sciences
Ravensburg-Weingarten
Weingarten, Baden-Württemberg, Germany

ISSN 2198-4182   ISSN 2198-4190   (electronic)
Studies in Systems, Decision and Control
ISBN 978-3-030-12848-7   ISBN 978-3-030-12846-3   (eBook)
https://doi.org/10.1007/978-3-030-12846-3

Library of Congress Control Number: 2019930978

This Springer imprint is published by the registered company Springer Nature Switzerland AG
The registered company address is: Gewerbestrasse 11, 6330 Cham, Switzerland

To my wife and my children

Preface

This book comprises research carried out in the last years. My friend Marcin Witczak from the Polish University of Zielona Góra was probably the most important person with numerous helpful comments and contributions. Very important were also the researchers at the renowned Institute of Control and Computation Engineering at this University led by Józef Korbicz: Mariusz Buciakowski, Paweł Majdzik, Krzysztof Patan, Marcin Pazera and Bartłomiej Sulikowski. Notable is also my former colleague Andreas Paczynski, who initiated the contacts with Poland.

Very essential were also the contributions from and the discussions with the research assistants in my research projects: Marek Stania, Michał Zając and Paweł Ziemniak (project: *steering and braking systems for autonomous vehicles in production and service*), Anna Chami (project: *application of advanced control and diagnosis processes for pump systems*), Claudius Spindler and Piotr Witczak (project: *production energy efficiency prognosis system*) as well as Denis Hock and Johannes Schmelcher (project: *development of an innovative cabin bike*). Significant contributions result from the research work with the colleagues Nicolai Beisheim, Robert Bjekovic, Peter Hertkorn, Jens Kiefer, Michael Niedermeier, Stephan Rudolph and Markus Till as well as the Research Assistants Theresa Breckle, Michael Elwert, Kevin Holder, Manuel Ramsaier, Fabian Wünsch and Andreas Zech in the project *digital product life cycle*. Very positive were the discussions with my office colleague Wolfgang Engelhardt. A positive influence on my research work had the good climate at our department; important persons for this climate are (apart from the ones mentioned above) Zerrin Harth, Bernhard Bauer, Edmund Böhm, Thomas Glogowski, Jörg Hübler, Andre Kaufmann, Tim Nosper, Thomas Schreier-Alt and Michael Winkler. A part of the preparation of this book was an interview series in industry; Bernhard Freyer, Florian Engstler, Christian Mergl, Franz Müller, Udo Pulm and Markus Viertlböck were, amongst others, willing to be my interview partners.

I would like to thank all of them, the other employees at the Hochschule Ravensburg-Weingarten as well as the numerous students involved in the projects for their input, their openness, their understanding and the fun in the projects. I also thank the management of my university, the Hochschule Ravensburg-Weingarten, especially the rector Thomas Spägele for the support and the freedom. Peter Eckart took over some teaching obligations in my research semester; I am grateful for this.

I especially thank my sister Eva for the final correction of the book.

Finally, I thank my wife and children for the time they allowed me for finishing this book and my mother for the continuous support.

Ravensburg, Germany   Ralf Stetter
December 2018

Contents

Part II Fault-Tolerant Design and Control of Automated Vehicles

Part III Fault-Tolerant Design and Control of Automated Processes

Acronyms

AFD	Active Fault Diagnosis
AGV	Automatic Guided Vehicle
AI	Artificial Intelligence
AR	Augmented Reality
ARMA	Autoregressive Moving Average
BMC	Bayesian Monte Carlo
CAD	Computer Aided Design
CDF	Cumulative Distribution Function
CFD	Computational Fluid Dynamics
CRB	Cohesion of Rigid Bodies
COG	Centre Of Gravity
CRB	Cohesion of Rigid Bodies
DAE	Differential Algebraic Equations
DES	Discrete Event System
DfC	Design for Control
DfD	Design for Diagnosis
DfM	Design for Monitoring
DfS	Design for Safety
EKF	Extended Kalman Filter
EMC	Electromagnetic Compatibility
EoL	End-of-Life
ERP	Enterprise Resource Planning
FDD	Fault Detection and Diagnosis
FDI	Fault Detection and Isolation
FEM	Finite Element Method
FMEA	Failure Mode and Effects Analysis
FMS	Flexible Manufacturing System
FPGA	Field Programmable Gate Arrays
FSA	Finite State Automata
FT	Failure Threshold

FTA	Fault Tree Analysis
FTAP	Fault-Tolerant Approaches
FTC	Fault-Tolerant Control
FTD	Fault-Tolerant Design
FTMD	Fault-Tolerant Mechanism Design
FR	Functional Redundancy
GPS	Global Positioning System
HI	Health Indicator
HS	Health Stage
IBP	Internal Battery Parameter
IC	Integrated Circuit
ICR	Instantaneous Centre of Rotation
LHS	Left Hand Side
LIDAR	Light Detection and Ranging
LMI	Linear Matrix Inequality
LPV	Linear Parameter-Varying
LUL	Loading/UnLoading (system)
LW-PLS	Locally-Weighted Partial Least Squares
LSSVM	Least Square Support Vector Machine
MD	Mechanism Design
MES	Manufacturing Execution System
MBS	Multi-Body System
MPC	Model Predictive Control
MSO	Minimal Structurally Overdetermined
MTTF	Mean-Time-to-Failure
MTTR	Mean-Time-to-Repair
NVH	Noise Vibration Harshness
ODE	Ordinary Differential Equations
OHC	Overhead Camshaft (Engine)
OHV	Overhead Valve (Engine)
ORKF	Outlier-Robust Kalman Filter
PDE	Partial Differential Equations
PDF	Probability Density Function
PDM	Product Data Management
PFC	Product Function Correntropy
PFD	Passive Fault Diagnosis
PMDD	Product Model Driven Development
RHS	Right Hand Side
RVM	Relevance Vector Machine
RUL	Remaining Useful Life
RM	Requirements Management
SED	Sensor Error Detection
SOC	State of Charge
SOH	State of Health
SVM	Support Vector Machine

SysML	System Modelling Language
TMR	Triple Modular Redundancy
TIPS	Theory of Inventive Problem Solving
UKF	Unscented Kalman Filter
UML	Unified Modelling Language
VR	Virtual Reality

Symbols

t	Time
k	Discrete time
$x_k, \hat{x}_k \in \mathbb{R}^n$	State vector and its estimate
$y_k, \hat{y}_k \in \mathbb{R}^m$	Output vector and its estimate
$y_{M,k} \in \mathbb{R}^m$	Model output
$u_k \in \mathbb{R}^r$	Input vector
w_k	Disturbance vector
$f_k \in \mathbb{R}^s$	Fault vector
$f(\cdot), g(\cdot), h(\cdot)$	Non-linear functions
α_k	Gain degradation
Z	Variable
K	Known variable
G	Structure graph
L	Distribution matrix
M	Incidence matrix
A, B, C	System matrices
J	Cost function
N_p	Prediction horizon
CCV	Customer comparative value
W_P	Weight of the performance
P	Assessment of the expected performance
W_A	Weight of the availability
A	Assessment of the expected availability
W_{IC}	Weight of the investment costs
IC	Assessment of the expected investment costs
W_{OC}	Weight of the operating costs
OC	Assessment of the expected operating costs
W_{vel}	Weight of the velocity
vel	Assessment of the expected velocity

W_{acc}	Weight of the acceleration
acc	Assessment of the expected acceleration
W_M	Weight of the manoeuvrability
M	Assessment of the expected manoeuvrability
W_{CC}	Weight of the carrying capacity
CC	Assessment of the expected carrying capacity
W_{RR}	Weight of the room requirements
RR	Assessment of the expected room requirements
W_{EC}	Weight of the energy costs
EC	Assessment of the expected energy costs
W_{SC}	Weight of the service costs
SC	Assessment of the expected service costs
W_{SurC}	Weight of the surveillance costs
Sur	Assessment of the expected surveillance costs
COG	Centre of gravity
$i = front, rear$	Axle location
$j = left, right$	Wheel location
L_f	Distance between front axle and COG
L_r	Distance between rear axle and COG
L_a	Rear/front half gauge
r	Yaw rate
v_x	Longitudinal velocity
a_x	Longitudinal acceleration
a_y	Lateral acceleration
m	Mass
δ_f	Steering angle of the front wheels
δ_r	Steering angle of the rear wheels
F_x	Sum of forces causing longitudinal motion
F_y	Sum of forces causing lateral motion
$F_{x,ij}$	Longitudinal force on i,j wheel
$F_{y,f}$	Total lateral force on the front wheels
$F_{y,r}$	Total lateral force on the rear wheels
$\omega_{i,j}$	Angular velocity of i,j wheel
T	Total torque acting on all wheels
$p_{i,j}$	Torque distribution coefficient
I_{xw}	Wheel moment of inertia
I_z	Robot moment of inertia around z-axis
R_e	Wheel effective radius
C_f	Front wheel cornering stiffness
C_r	Rear wheel cornering stiffness
DI_{P_0}	Performance oriented degradation indicator
DI_{C_0}	Consumption oriented degradation indicator
DI_{AO_0}	Additional output oriented degradation indicator
DI_{AS_0}	Auxiliary substance oriented degradation indicator

DI_{NVH_0}	Noise, vibration and harshness oriented degradation indicator
$P(c_i)$	Probability of the i-th cut set
$P(e_j)$	Probability of the e_j basic event
n	Number of basic events in the i-th cut set
P_T	Probability of the top event
m	Number of cut sets
λ_{R_E}	Ageing parameter for the electrolyte resistance
R_E	Electrolyte resistance
$\lambda_{R_{CT}}$	Ageing parameter for the charge transfer resistance
R_{CT}	Charge transfer resistance
C_I	Capacity at rated current
mpv_{IBP}	Model predicted value of an internal battery parameter
R_{Dif}	Diffusion resistance
C_{CT}	Charge transfer capacity
C_{Dif}	Diffusion capacity
τ_{CT}	Time constant charge transfer
τ_{Dif}	Time constant diffusion
$V(k)$	Average output voltage
$V_{CT}(k)$	Charge transfer voltage
$V_{Dif}(k)$	Diffusion voltage
V_{OCV}	Open circuit voltage
$f_i(k)$	Travel duration of an AGV on a journey in forward direction
$b_i(k)$	Travel duration of an AGV on a journey in backward direction
$v_i(k)$	Switching variable
$z(k)$	Auxiliary binary variable
k_f	Number of cycles from full state of charge down to zero charge
k_{f1}	Feasible number of cycles
k_{f2}	Remaining number of cycles
q	Weighting constant
β	Importance reflecting constant
R_i	ith resource
$d_i(k)$	Operation duration
$\bar{x}_i(k)$	Transportation start time
$\bar{y}(k)$	Storage zone seat availability time
$\bar{v}(k)$	Loading zone seat availability time
κ	Variable representing the importance of the state of charge

Chapter 1
Introduction

In the last decades, a continuous trend towards more complicated and complex technical systems can be observed. The main causes are increasing requirements from consumers and society as well as an intensified global competition and an instant growth of modern industry. In spite of the undeniable advantages of these complicated and complex systems in terms of functionality, efficiency, reliability and safety, the possibility and frequency of faults and failures increase. Consequently, high availability and reliability of such systems require some kind of fault-tolerance and necessitate continuous research activities in this direction. In this context, the essence of fault-tolerance is the capability of a technical system to accommodate one or more fault(s) and to bring the performance back within the region of acceptable performance.

Additionally, prominent trends underline the increasing importance of fault-tolerance because they also lead to new fault possibilities. The upcoming convergence of technologies, the exponentially growing functionality of products (especially in the fields comfort, entertainment and safety), the horizontal and vertical connectivity, the agile modular production concepts, the distributed development and production processes, the decreasing lot sizes (up to costumer specific products with lot size one), the intensified man-machine-cooperation and the overwhelming data abundance result in an increasing complexity of products and processes and impair the transparency of products and processes. Conventional means to increase the fault-tolerance, for instance redundancy, do not allow to address the increased number of fault possibilities without considerable disadvantages, for instance increasing costs and weight. An intensified scientific discourse concerning approaches, which can enhance the fault-tolerance without negative effects, is consequently a cornerstone for the successful development of intelligent technical systems.

This necessity has motivated researchers worldwide to concentrate on "Fault-Tolerant Control" (FTC), which deals with the control of systems that exhibit faults during their operating life and which is primarily meant to provide safety, that is, the stability of a system after the occurrence of a fault in the system (compare [5]). Recent research initiatives have shown that the design of technical systems

© Springer Nature Switzerland AG 2020
R. Stetter, *Fault-Tolerant Design and Control of Automated Vehicles and Processes*, Studies in Systems, Decision and Control 201,
https://doi.org/10.1007/978-3-030-12846-3_1

can support or aggravate control and diagnosis processes which are both a part of FTC. Consequently, research into the field of "Fault-Tolerant Design" (FTD) can pave the road to technical systems with integrated fault-tolerance. In this sense, FTD is understood as a collection of strategies, methods, algorithms, tools and insights which can support the development of technical systems which are fault-tolerant because of their controllability but also their inherent fault-tolerant design qualities. It is important to point out that fault-tolerant design can ease fault-tolerant control—the combination of both is extremely fruitful.

A main part of this book is devoted to method development, research application and system validation in the field of automatic vehicles and processes. It is important to note that the notion "automatic vehicles" includes vehicles in production or infrastructure scenarios but also passenger and transportation vehicles which are often referred to as "autonomous vehicles" or "Automated Guided Vehicles" (AGVs). In the main focus of "automatic processes" are the manufacturing, assembly and logistic processes in industrial companies, though similar processes may be present in infrastructure systems and processing plants. In recent years the notions "mechatronic products", "intelligent products", "industry 4.0" and "cyber-physical systems" were used for technical systems with an increasing amount of sensing, communication, adaptation and reconfiguration capability in distributed scenarios. In this book, usually the notion technical systems is used, but is including these kinds of intelligent systems.

The implementation of fault-tolerant control and fault-tolerant design requires certain processes, algorithms, strategies, methods and tools. Engineers are frequently challenged in the planning of the product development process of complex technical systems; especially when the priority of development tasks has to be decided. A conscious use of tools such as process models or sequence portfolios can help in this endeavour, but they have to be applied at the right time in the right manner. Requirements and their management are central success factors in product development. A central issue is the enormous number of requirements for complex technical systems. A very promising approach is using interconnected models or the product and its requirements in order to allow an improved collection, documentation and tracking of requirements. From a technical point of view, the availability of reliable sensor data is a frequent challenge. This problematic issue can at least partly be addressed by virtual sensors. Increasing requirements concerning the availability and reliability directly lead to demands of improved fault-tolerance. On the one hand, predictive fault-tolerant control can compensate the effect of faults and thus lead to increasing availability and reliability. On the other hand, the demands for improved fault-tolerance can be addressed, when it is possible to use a prognosis of the degradation behaviour of a system. Such approaches also need to include systems with redundancies and resource conflicts. The processes, algorithms, strategies, methods and tools listed have in common that they can support fault-tolerant design and fault-tolerant control and that they can contribute to improving the general fault-tolerance, but that they require a deep understanding of fault-tolerant design and fault-tolerant control.

1.1 Overview of Main Research Activities

This section describes the state-of-the-art in fault-tolerant control and design.

1.1.1 Fault-Tolerant Control

Fault-Tolerant Control (FTC) is an area of research that aims to increase availability and safety by specifically designing control algorithms which are capable of maintaining stability and performance regardless of the occurrence of faults and it has drawn considerable attention from the control research community and the engineering community in the past couple of decades [26]. The challenging efficiency and reliability requirements in many sectors require a continuous development of control and fault diagnosis processes and systems [6, 9, 13, 28, 30, 72] in theory and practice. Frequently, the requirements strech beyond the normally accepted safety-critical systems, which control, for instance nuclear reactors, chemical plants or aircrafts, to innovative systems such as distributed production systems, autonomous vehicles or high-speed rail systems. Since several decades, intensive research activities which address these issues have been centred on the notion "Fault Detection and Identification" (FDI). In the early stage of research, a large amount of attention was devoted to techniques for the development of advanced FDI [9] including the analytical and knowledge based approaches as well as computational intelligence techniques [27, 66, 71]. An historical overview of the development of FDI technology is given in Fig. 1.1.

The state-of-the-art in the analytical approaches based FDI is marked by the completed establishment of a framework of FDI theory for linear time-invariant systems and a current focus on FDI for non-linear and uncertain control systems [47, 71, 72]. In the areas of knowledge-based and computational intelligence techniques, the main trends are the integration of different FDI schemes aiming at improving the fault diagnosis system performance. Comparisons between parts of real complex systems and their computer-based counterparts designed with the knowledge of their mathematical models allow generating diagnostics signals, which are a vast information source.

Connected with the rapidly increasing requests for higher system performance, product quality, productivity and cost efficiency, fault diagnosis and FTC design become crucial concerns in product development and system design [29, 30, 36, 38, 45, 46, 64, 73] and therefore receive much attention in the industry and the academic community. Generally, FTC schemes can be divided into two categories [6, 73]: passive and active FTC; this distinction will be explained in detail in Chap. 2.

The conventional way to increase the system fault-tolerance consists in building hardware redundancy (also called backup) [23] for key system components like sensors, actuators, computers and bus systems. For a part of system components, this technique has been expanded in the last years to include software redundancy.

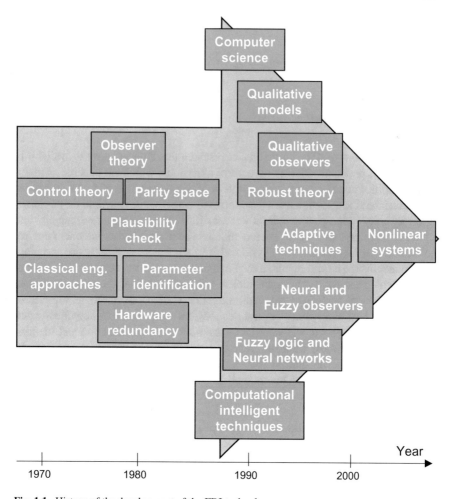

Fig. 1.1 History of the development of the FDI technology

Prominent research approaches also concern reconfiguration schemes [6, 72]. Recently, "Linear Parameter-Varying" (LPV) systems [12, 51, 52] receive a large research attention in scope of FDI and FTC for non-linear systems, due to their ability to decently reflect the behaviour of non-linear systems. This ability combined with their analytically appealing structure makes them perfect candidates for the representation of non-linear dynamic subsystems.

1.1.2 Fault-Tolerant Design

As stated above, fault-tolerant design (FTD) is understood as a collection of strategies, methods, algorithms, tools and insights which can support the development

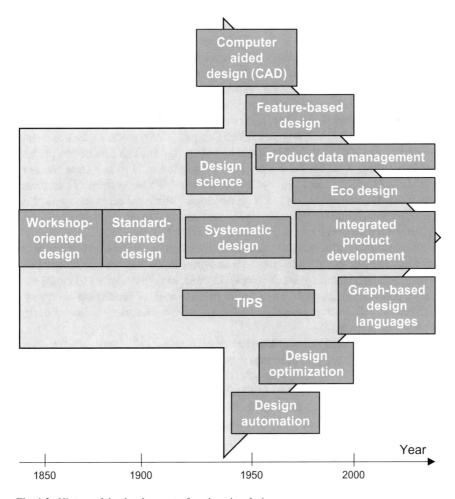

Fig. 1.2 History of the development of engineering design

of technical systems which are fault-tolerant because of their controllability but also their inherent fault-tolerant design qualities. In spite of the numerous innovative and challenging research directions which are currently developing in FTC, it can be concluded that this field is intensively researched and a wide collection of excellent research results and applicable algorithms and systems are available. In contrast, research concerning Fault-Tolerant Design (FTD) is sparse and an orientation in this field is challenging. There is a rich body of research concerning systematic design and the strategies, methods and tools of product development [3, 11, 14, 17, 21, 35, 41, 43, 50, 63, 67], which can present an excellent basis for the development of a FTD methodology; FTD was however usually not a point of main emphasis. An historical overview of the development of engineering design is given in Fig. 1.2.

In this context, the "Theory of Inventive Problem Solving" (TIPS—a collection of tools for problem solving and innovation based on the study of patterns in patent literature by G. Altshuller) may expand this basis. Research concerning the general topic of design automation is also notable (for a general framework compare [10]). An integrated product and process evaluation [65] are in the scope of current work. Connected research is carried out in the field of design optimization [18]; major research activities concern topology optimization, which is a computational technique for distributing material efficiently in design spaces in order to optimise stiffness and reduce mass [4, 48, 54]. Current research in feature based design (i.e. the application of intelligent "Computer Aided Design" (CAD) models which associate geometrical knowledge and other kinds of knowledge such as production process knowledge [15]) concerns for instance the integration in "Product Data Management" (PDM) and "Product Life-cycle Management" (PLM) systems [7]. In recent years a new research direction has been developed which concerns the application of graph-based design languages in design and can lead to an automated execution of holistic product development processes [19, 48, 49].

In the last decades the general topic of sustainability has received rising attention. This trend has led to enormous research activities in the field of eco-design [37]. Several tools and systems were developed in order to help designers to create sustainable products, for instance a system for the prognosis of production energy in a CAD system [61]. Further research is focused on the evolutionary nature of design processes in industry [58].

In the scientific field of control engineering, Isermann [22], amongst others, also addresses system development issues; these research results and proposals can also serve as a basis for the development of FTD.

An overview of research results from reliability engineering is given by O'Connor and Kleyner [39]. Current research in this field deals with fuzzy assessing [24], dynamic Bayesian networks [33] and structural reliability analysis with multiple imprecise random and interval fields [16]. The most important safety standards in this area are EN ISO 13849-1 [2] and EN 62061 [1].

It is also possible to identify research activities which are directly concerned with FTD, but usually they are limited to very specific fields of application. Rouissi and Hoblos [53] remark that the ability of a system (sensors, actuators, process) to tolerate one or more fault(s) must be achieved by means of designing and must focus on a sensor network design. Oh et al. [40] aim at fault-tolerant design approaches for nuclear power plants and include several fault-tolerance and fault avoidance design features such as the 2-out-of-4 (2/4) voting logic, redundant actuation devices and certain requirements towards process instrumentation. Lin and Yang [34] use the term "fault-tolerant design" for the design of reliable wide area mobile networks. Shirazipourazad et al. [55] report about FTD of wireless sensor networks and concentrate on approximation algorithms aimed at enhancing the fault tolerance of a tree gathering data from directional antennas. In the field of microelectronics Hsieh et al. [20] list "Triple Modular Redundancy" (TMR—two additional copies of a certain target design are added in the chip and a voter is used—in this case the chip can work correctly if at least two of the copies can work) and error-tolerance (chips with minor errors that are not visible to human end users) as examples of fault-tolerant design approaches for chip architecture. Vedachalam et al. [68] propose a

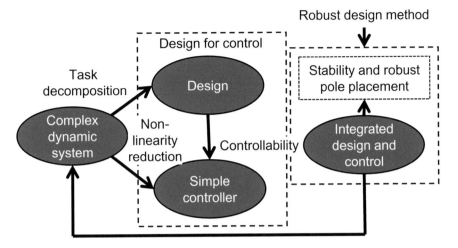

Fig. 1.3 Integrated design and control methodology

fault-tolerant design approach for reliable frequency converters. They apply the methods "reliability modelling" and "redundancy analysis" in order to enable an optimum redundancy implementation in order to decide the trade-off between the reliability, performance, cost, and size of such converters. Straka et al. [62] describe fault tolerant system design approaches for "Field Programmable Gate Arrays" (FPGA) and focus on fault recovery capabilities using reconfigurable modules and interfaces. A similar notion—"Fault Tolerant Mechanism Design (FTMD)" is proposed by Porter et al. [44] in the field of artificial intelligence. In this scope mechanism design (MD) is a sub-area of game theory and is the science of crafting protocols for self-interested agents. The difference between MD and FTMD is that FTMD agents have richer private information (additionally probability of failure). It is obvious that their research is in a different field than this book. However, the general approach to apply individual units (agents) with higher intelligence allows analogical considerations in the field of fault-tolerant design of technical systems. Similarly, the research work of Jiang and Hsu [25] in the field of cloud logistics allows a methodological analogy when looking at the application of "Petri nets" for fault-tolerant design.

As FTD includes elements of control and diagnosis, also the research approaches aiming at "Design for Control" (DfC) and "Design for Diagnosis" (DfD) can be considered. One approach for DfC was presented by Li et al. [32]; however, the scope was limited to the synthesis of simple dynamic models of the mechanical structures and controller design. Related work is described by Li and Lu [31] which formulate initial abstract schemes for an integrated design and control methodology (Fig. 1.3). An in-depth investigation of approaches how technical systems can be designed in order to facilitate and to simplify effective and efficient control is reported by Stetter and Simundsson [60].

Research towards DfD tries to answer the question how products should be designed in order to allow and ease diagnosis, i.e. the detection and identification of product or process abnormalities. In the scope of developing complex electronic systems initial approaches to face this challenge are reported. Chen [8] uses the notion Design for Diagnosis for describing a technique to measure the delay and crosstalk noise for the testing and the diagnosis of on-chip bus wires and Wang et al. [69] introduce a technique for diagnosing faults within "Integrated Circuits" (IC). General valid guidelines for DfD were formulated by Stetter and Phleps [59]. Guidelines for "Design for Monitoring" (DfM [57]) can be understood as an intermediate step.

It can be concluded that a number of research activities have already started in the general scope of fault-tolerant design and that these activities have been intensified in the last years. However, until now these activities are not connected and no logical and scientific abstract classification of these activities has been generated so far.

1.2   Essential Concepts of Fault-Tolerance

One ultimate goal of fault-tolerance is an increase of the safety of technical systems—consequently safety is the essential concept of fault-tolerance. The prominent safety standards EN ISO 13849-1 [2] and EN 62061 [1] define safety as freedom from unacceptable risk. The main objective of safety is to protect human beings from injury and death and technical systems and the environment from any kind of damage. A central role in industry play the Machinery Directive 2006/42/EC and harmonised standards listed under it. EN ISO 13849-1 [2] describes general principles for the design for safety-related parts of control system. The main emphasis of EN 62061 [1] is the functional safety of safety-related electrical, electronic and programmable electronic control systems. It is the decision of industrial manufacturers, which one of the two safety system standards they use. In any case, industrial manufactures have to design their technical systems in a manner that they correspond with the state of the art in safety-related fields and have to provide safety functions by detecting dangerous conditions and by bringing all operations and processes to a safe state.

Additionally, two essential concepts are the *reliability* and the *availability*. Reliability is the probability that a system will perform its intended function for a specified period of time (design life) under specified design operating conditions (e.g. load, ambient temperature) [42]. The two important measures often used in maintenance studies "Mean-Time-To-Failure" (MTTF—denotes the average time to failure) and "Mean-Time-To-Repair" (MTTR—denotes the mean downtime or the expected value of the repair time) help to clarify the availability of a system:

$$Availability = \frac{System\ up\ time}{System\ up\ time + System\ down\ time} = \frac{MTTF}{MTTF + MTTR} \quad (1.1)$$

Obviously, the availability of technical systems is closely connected to maintenance. Apart from control and diagnostic actions, maintenance costs constitute

Fig. 1.4 Hierarchy of systems

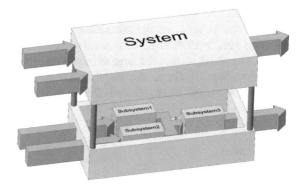

a large portion of the operating and overhead expenses in complex systems [70]. A large portion of this cost results from inefficient maintenance operations, such as unscheduled down-times due to unexpected failures, poor spare parts logistics, and replacement policies. Another reason of such an unappealing effect is related to an inappropriate exploitation of system components, which leads them to the premature failure. Thus, recent research focuses on estimating the "Remaining Useful Life" (RUL) [56] of the system components. Subsequently, RUL can be used for predicting the "Mean-Time-To-Failure" (MTTF) of the system components as well as to control and maintain the system in such a way to enhance this time.

Current technical systems dispose of a modular structure and are realised through a hierarchy of subsystems. In general, complex systems are composed of simpler subsystems (compare Fig. 1.4) and can only function, if the subsystems work together in an organised manner.

The general concept of aggregating subsystems to complex systems has consequences on many aspects of control and diagnosis. Some prominent aspects of the consequences of the hierarchy of subsystems and its consequences to prognostics are discussed in Sect. 2.2.

1.3 Structure of this Book

The next two chapters aim to deliver the fundamental aspects and principles of fault-tolerance. Chapter 2 concentrates on fault-tolerant control, while Chap. 3 discusses the relatively new field of fault-tolerant design. The two later parts build upon this basis and describe concrete tools, algorithms, methods, frameworks, strategies and systems for both aspects. The second part of this book is concerned with fault-tolerant control and design of automated vehicles. The processes, methods and tools which allow a methodical and model-based design of automated vehicles are discussed in Chap. 4. Chapter 5 proposes an approach for implementing virtual sensors. This chapter concerns a continuous system description, whereas the later chapters will focus on "Discrete Event Systems" (DES). The third part of the book deals with the fault-tolerant design and control of automated processes. Chapter 6 explains a framework for the fault-tolerant control of the automated processes in a complex assembly system and Chap. 7 estimation strategies for the "Remaining Useful Life" (RUL). Chapter 8 extends the topic of predictive fault-tolerant control towards automated processes which can dispose of both flexible redundant and shared elements. The conclusions and future research directions are given in Chap. 9.

References

1. EN 62061:2005 (Safety of Machinery—functional Safety of safety-related Electrical, Electronic and Programmable Electronic Control Systems)
2. EN ISO 13849-1:2008 (Safety of machinery—safety-related parts of control system—general principles for design)
3. Andreasen, M.M., Hansen, T.C., Cash, C.P.: Conceptual Design. Mindset and Models. Springer, Interpretations (2015)
4. Banh, T.T., Lee, D.: Multi-material topology optimization design for continuum structures with crack patterns. Compos, Struct (2017)
5. Benosman, M.: A survey of some recent results on nonlinear fault tolerant control. Math. Probl. Eng. **586169**, (2010)
6. Blanke, M., Kinnaert, M., Lunze, J., Staroswiecki, M.: Diagnosis and Fault-Tolerant Control. Springer, New York (2016)
7. Camba, J.D., Contero, M., Companyc, P., Prez, D.: On the integration of model-based feature information in product lifecycle management systems. Int. J. Informat. Manag. **37**, 611–621 (2017)
8. Chen, G.-N.: A Design for Diagnosis Technique for the Delay and Crosstalk Measurement of On-Chip Bus Wires. National Central University, Chung-Li, Taiwan (2000)
9. Chen, J., Patton, R.J.: Robust Model Based Fault Diagnosis for Dynamic Systems. Kluwer Academic Publishers, London (1999)
10. Chung, J.C.H., Hwang, T.-S., Wu, C.T., Jiang, C.-T., Wang, J.-Y., Bai, Y., Zou, H.: Framework for integrated mechanical design automation. Comput.-Aided Design **32**, 355–365 (2000)
11. Cross, N.: Engineering Design Methods: Strategies for Product Design. John Wiley and Sons Ltd. (2008)
12. de Oca, S., Puig, V., Witczak, M., Dziekan, L.: Fault-tolerant control strategy for actuator faults using LPV techniques: application to a two degree of freedom helicopter. Int. J. Appl. Math. Comput. Sci. **22**(1), 161–171 (2012)
13. Ding, S.X.: Model-based Fault Diagnosis Techniques: Design Schemes. Algorithms and Tools. Springer, Berlin (2008)
14. Ehrlenspiel, K., Meerkamm, H.: Integrierte Produktentwicklung. Zusammenarbeit. Carl Hanser Verlag, Denkabläufe, Methodeneinsatz (2013)
15. Fougeres, A.-J., Ostrosi, E.: Intelligent agents for feature modelling in computer aided design. J. Computat, Design Eng (2017)
16. Gao, W., Wu, D., Gao, K., Chen, X., Tin-Loi, F.: Structural reliability analysis with imprecise random and interval fields. Appl. Math. Modell. **55**, 49–67 (2018)
17. Hales, C., Gooch, S.: Managing Engineering Design. Springer Science and Business Media (2004)
18. Herrema, A.J., Wiese, N.M., Darling, C.N., Ganapathysubramaniana, B., Krishnamurthya, A., Hsua, M.-C.: A framework for parametric design optimization using isogeometric analysis. J. Comput. Methods Appl. Mech. Eng. **316**, 944–965 (2017)
19. Holder, K., Zech, A., Ramsaier, M., Stetter, R., Niedermeier, H.-P., Rudolph, S., Till, M.: Model-based requirements management in gear systems design based on graph-based design languages. Appl. Sci. **7**, (2017)

20. Hsieh, T.-Y., Li, K.-H., Chung, C.-C.: A fault-analysis oriented re-design and cost-effectiveness evaluation methodology for error tolerant applications. Microelectron. J. **66**, 48–57 (2017)
21. Hubka, V., Eder, W.E.: Theory of Technical Systems: A Total Concept Theory for Engineering Design. Springer (1988)
22. Isermann, R.: Fault Diagnosis Systems. An Introduction from Fault Detection to Fault Tolerance. Springer, New York (2006)
23. Isermann, R.: Fault Diagnosis Applications: Model Based Condition Monitoring, Actuators, Drives, Machinery, Plants, Sensors, and Fault-tolerant Systems. Springer, Berlin (2011)
24. Jamali, S., Bani, M.J.: Application of fuzzy assessing for reliability decision making. In: Proceedings of the World Congress on Engineering and Computer Science (2017)
25. Jiang, F.-C., Hsu, C.-H.: Fault-tolerant system design on cloud logistics by greener standbys deployment with petri net model. Neurocomputing **256**, 90–100 (2017)
26. Jiang, Y., Qinglei, H., Ma, G.: Adaptive backstepping fault-tolerant control for flexible spacecraft with unknown bounded disturbances and actuator failures. ISA Trans. **49**(1), 57–69 (2010)
27. Korbicz, J., Kościelny, J., Kowalczuk, Z., Cholewa, W. (Eds.).: Fault Diagnosis. Models, Artificial Intelligence, Applications. Springer, Berlin (2004)
28. Kościelny, J.M.: Diagnostics of Automatic Industrial Processes. Academic Publishers, Office EXIT (2001)
29. Kowalczuk, Z., Olinski, K.E.: Sub-optimal fault-tolerant control by means of discrete optimization. Int. J. Appl. Math. Comput. Sci. **18**(4), 50–61 (2008)
30. Lee, Y.I., Cannon, M., Kouvaritakis, B.: Extended invariance and its use in model predictive control. Automatica **41**(12), 2163–2169 (2005)
31. Li, H.-X., Lu, X.: System Design and Control Integration for Advanced Manufacturing. Wiley and Sons Ltd., Zurich (2015)
32. Li, Q., Zhang, W.J., Chen, L.: Design for control—a concurrent engineering approach for mechatronic systems design. IEEE/ASME Trans. Mechatron. **6**, 161–169 (2001)
33. Liang, X.F., Wang, H.D., Yi, H., Li, D.: Warship reliability evaluation based on dynamic bayesian networks and numerical simulation. Ocean Eng. **136**, 129–140 (2017)
34. Lin, J.-W., Yang, M.-F.: Fault-tolerant design for wide-area mobile ipv6 networks. J. Syst. Softw. **82**, 1434–1446 (2009)
35. Lindemann, U.: Methodische Entwicklung technischer Produkte. Springer (2009)
36. Liu, M., Cao, X., Shi, P.: Fault estimation and tolerant control for fuzzy stochastic systems. Trans. Fuzzy Syst. **21**(2), 221–229 (2013)
37. MacDonald, E.F., She, J.: Seven cognitive concepts for successful eco-design. J. Clean. Product. **92**, 23–36 (2015)
38. Noura, H., Sauter, D., Hamelin, F., Theilliol, D.: Fault-tolerant control in dynamic systems: application to a winding machine. IEEE Control Syst. Magaz. **20**(1), 33–49 (2000)
39. O'Connor, P.D.T., Kleyner, A.: Practical Reliability Engineering. John Wiley and Sons, Ltd (2012)
40. Oh, Y.G., Jeong, J.K., Lee, J.J., Lee, Y.H., Baek, S.M., Lee, S.J.: Fault-tolerant design for advanced diverse protection system. Nucl. Eng. Technol. **45**(6), 795–802 (2013)
41. Pahl, G., Beitz, W., Feldhusen, J., Grote, K.H.: Engineering Design: A Systematic Approach. Springer (2007)
42. Pham, H.: System Software Reliability. Springer (2006)
43. Ponn, J., Lindemann, U.: Konzeptentwicklung und Gestaltung technischer Produkte. Springer (2011)
44. Porter, R., Ronen, A., Shoham, Y., Tennenholtz, M.: Fault tolerant mechanism design. Artif. Intell. **45**(6), 1783–1799 (2013)
45. Pourmohammad, S., Fekih, A.: Fault-tolerant control of wind turbine systems—a review. In: Proceedings of the Green Technologies Conference (IEEE-Green), pp. 1–6 (2011)
46. Rafajlowicz, E., Rafajlowicz, W.: Control of linear extended nd systems with minimized sensitivity to parameter uncertainties. Multidimens. Syst. Signal Process. **24**(4), 637–656 (2013)
47. Rafajlowicz, E., Styczen, K., Rafajlowicz, W.: A modified filter sqp method as a tool for optimal control of nonlinear systems with spatio-temporal dynamics. Int. J. Appl. Math. Comput. Sci. **22**(2), 313–326 (2012)

48. Ramsaier, M., Spindler, C., Stetter, R., Rudolph, S., Till, M.: Digital representation in multi-copter design along the product life-cycle. Procedia CIRP. **62**, 559–564 (2016)
49. Ramsaier, M., Stetter, R., Till, M., Rudolph, S., Schumacher, A.: Automatic definition of density-driven topology optimization with graph-based design languages. In: Proceedings of the 12th World Congress on Structural and Multidisciplinary Optimisation (2017)
50. Roozenburg, N.F.M., Eekels, J.: Product Design: Fundamentals and Methods. Wiley (1995)
51. Rotondo, D., Puig, V., Nejjari, F., Romera, J.A.: A modified filter sqp method as a tool for optimal control of nonlinear systems with spatio-temporal dynamics. Quasi-Lpv Fault-Toler. Control. Four-Wheel. Omnidirectional Mob. Rob. **62**(6), 3932–3944 (2015)
52. Rotondo, D., Puig, V., Nejjari, F., Witczak, M.: Automated generation and comparison of Takagi-Sugeno and polytopic quasi-LPV models. Fuzzy Sets Syst. **277**(C), 44–64 (2015)
53. Rouissi, F., Hoblos, G.: Fault tolerant sensor network design with respect to diagnosability properties. In: Proceedings of the 8th IFAC Symposium on Fault Detection, Supervision and Safety of Technical Processes (SAFEPROCESS), pp. 1120–1124 (2012)
54. Schmelcher, J., Stetter, R., Till, M.: Integrating the ability for topology optimization in a commercial cad-system. In: Proceedings of the 20th International Conference on Engineering Design (ICED 15), Vol 8: Vol 6: Design Methods and Tools Part 2, pp. 173–182 (2015)
55. Shirazipourazad, S., Sen, A., Bandyopadhyay, S.: Fault-tolerant design of wireless sensor networks with directional antennas. Pervas. Mobile Comput. **13**, 258–271 (2014)
56. Si, X.-S., Wang, W., Hu, C.-H., Zhou, D.-H.: Remaining useful life estimationa review on the statistical data driven approaches. Eur. J. Oper. Res. **213**(1), 1–14 (2011)
57. Stetter, R.: Monitoring in product development. In: Conference Proceedings of the 14th European Workshop on Advanced Control and Diagnosis (ACD) (2017)
58. Stetter, R., Möhringer, S., Günther, J., Pulm, U.: Investigation and support of evolutionary design. In: Proceedings of the 20th International Conference on Engineering Design (ICED 15) Vol 8: Innovation and Creativity, pp. 183–192 (2015)
59. Stetter, R., Seemüller, H., Chami, M., Voos H.: Interdisciplinary system model for agent-supported mechatronic design. In: Proceedings of the 18th International Conference on Engineering Design (ICED11)
60. Stetter, R., Simundsson, A.: Design for control. In: Proceedings of the 21st International Conference on Engineering Design (ICED 17) Vol 4: Design Methods and Tools, pp. 149–158 (2017)
61. Stetter, R., Witczak, P., Witczak, Kauf, F., Staiger, B., Spindler, C.: Development of a system for production energy prognosis. In: Proceedings of the 20th International Conference on Engineering Design (ICED 15) Vol 1: Design for Life, pp. 107–116 (2015)
62. Straka, M., Kastil, J., Kotasek, Z., Miculka, L.: Fault tolerant system design and seu injection based testing. Microprocess. Microsyst. **37**, 155–173 (2013)
63. Suh, N.P.: Konzeptentwicklung und Gestaltung technischer Produkte. Oxford University Press (2001)
64. Tatjewski, P.: Advanced Control of Industrial Processes: Structures and Algorithms. Advances in Industrial Control. Springer, London (2007)
65. Tornowa, A., Graubohm, R., Dietrich, F., Drder, K.: Design automation for battery system variants of electric vehicles with integrated product and process evaluation. Procedia CIRP **50**, 424–429 (2016)
66. Trave-Massuyes, L.: Bridging control and artificial intelligence theories for diagnosis: a survey. Eng. Appl. Artif. Intell. **27**, 1–16 (2004)
67. Ulrich, K.T., Eppinger, S.D.: Product Design and Development. McGraw-Hill (2008)
68. Vedachalam, N., Umapathy, A., Ramadass, G.A.: Fault-tolerant design approach for reliable offshore multi-megawatt variable frequency converters. J. Ocean. Eng. Sci. **1**, 226–237 (2016)
69. Wang, F., Hu, Y., Li, X.: A design-for-diagnosis technique for diagnosing integrated circuit faults with faulty scan chains. In: Proceedings of the IEEE 8th Workshop on RTL and High Level Testing (2007)
70. Wang, H.: A survey of maintenance policies of deteriorating systems. Eur. J. Oper. Res. **139**(3), 469–489 (2002)

71. Witczak, M.: Modelling and Estimation Strategies for Fault Diagnosis of Non-linear Systems. Springer, Berlin (2007)
72. Witczak, M.: Fault Diagnosis and Fault-Tolerant Control Strategies for Non-Linear Systems. Springer, Analytical and Soft Computing Approaches (2014)
73. Zhang, Y., Jiang, J.: Bibliographical review on reconfigurable fault-tolerant control systems. Ann. Rev. Control **32**(2), 229–252 (2008)

Part I
Principles of Fault-Tolerant Design and Control

Chapter 2
Fault-Tolerant Control

For the sake of discussing the main aspects, possibilities and necessities of fault-tolerant control, automatic vehicles and processes can both be seen as controlled, dynamic systems. In such systems, actuators, the plant itself and sensors can be distinguished. Figure 2.1 shows a general scheme of a controlled dynamic system.

Such systems can be described either in the continuous-time domain using the subsequent differential equations:

$$\dot{x}(t) = f(x(t), u(t)), \tag{2.1}$$
$$y(t) = h(x(t), u(t)), \tag{2.2}$$

or in the discrete-time domain using the subsequent recursive equations [49]:

$$x_{k+1} = f(x_k, u_k), \tag{2.3}$$
$$y_k = h(x_k, u_k), \tag{2.4}$$

where $x \in \mathbb{R}^n$ is the state vector, $y \in \mathbb{R}^m$ is the output vector, and $u \in \mathbb{R}^r$ denotes the control input vector. f and h are (in general) non-linear functions.

In real-life application, such systems are frequently affected by *faults*. A fault may generally be defined as an unpermitted deviation of at least one characteristic property or parameter of the system from the acceptable, usual, standard condition [27], e.g. an actuator malfunction. All the unexpected variations that have a tendency to degrade the overall performance of a system or its components can also be interpreted as faults [69].

The term *failure* describes permanent interruptions of the ability of a system or component to perform an intended function under specified operating conditions [27]. Usually the term "failure" suggests a complete breakdown of a system, whereas the term "fault" denotes a malfunction rather than a catastrophe [69]. It is important to distinguish faults from disturbances. Disturbances cannot be avoided in real-life applications, but their effects on the system performance can be attenuated by means

© Springer Nature Switzerland AG 2020
R. Stetter, *Fault-Tolerant Design and Control of Automated Vehicles and Processes*, Studies in Systems, Decision and Control 201,
https://doi.org/10.1007/978-3-030-12846-3_2

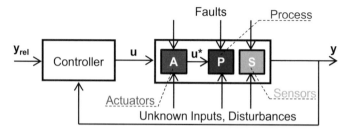

Fig. 2.1 General scheme of a controlled dynamic system

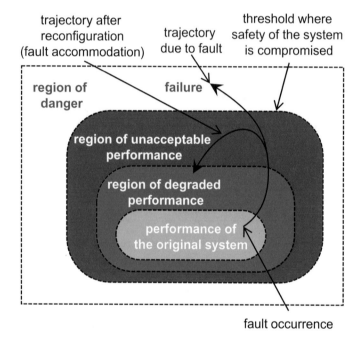

Fig. 2.2 General relationships between faults, failure, region of danger and fault accommodation

of carefully designed robust controllers or by means of measures such as filtering [45]. Faults typically lead to more drastic changes in the dynamics of plants, which result in effects that cannot be overcome by a fixed controller. The effect of faults may increase over time and can lead to a complete failure of the system, if no appropriate countermeasures are taken. Fault-tolerant control aims at accommodating the effects of the fault and at preventing failure. Figure 2.2 explains the relationships between faults, failure, region of danger and fault accommodation.

Figure 2.2 shows a typical evolution of the performance of a system simultaneously with an increase of the level of a fault in one of its subsystems. If any fault is absent, the system can operate in its normal region of performance. At the beginning, when the size of the fault is small, the usually present feedback control may obscure

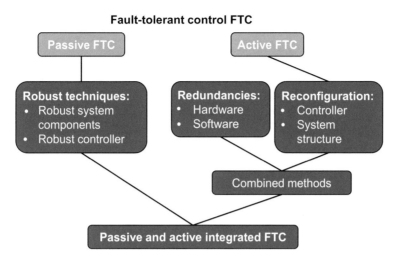

Fig. 2.3 Basic classification of FTC concepts

the presence of the fault. When the size of the fault increases, the performance of the system degrades. In the case that no other control measures are realised, the system can eventually become unstable [45]. The objective of fault-tolerant control is to accommodate the fault and to return the performance of the system into the region of acceptable performance.

Fault-tolerant control concepts can be divided into two categories [8, 74]: *passive* and *active* approaches. The crucial distinction between these two can be expressed by the fact that the active FTC system includes an FDI system and the fault handling is realised based on information on faults received from this FDI system, while in a passive FTC system, the system components and controllers are designed in a manner which enables them to be robust to possible faults up to a certain degree. Figure 2.3 presents the basic classification of FTC concepts.

In a passive fault-tolerant system, the controller does not react, if a fault occurs. The structure and the parameters of the controller are designed in a manner which facilitates the system to tolerate a set of faults without any active change [60]. In case of redundant actuators, the passive approach is also referred to as *reliable control methods* [75].

The distinctive quality of active approaches is that a new set of control parameters or a new structure of the control is applied in the faulty case such that the faulty system can still achieve the nominal system performance. In active fault-tolerant control systems, a fault detection and diagnosis (FDD) scheme can detect and diagnose the respective fault. After this diagnosis, the controller will be redesigned or even be reconfigured in the case of severe faults. The term "control reconfiguration" describes the problem of changing the control law or the controller structure by chosing an alternative set of inputs and outputs. After this choice of an alternate configuration, new control parameters should be developed which facilitate the new controller to be able to realise the original system performance, if it is possible, or, at least,

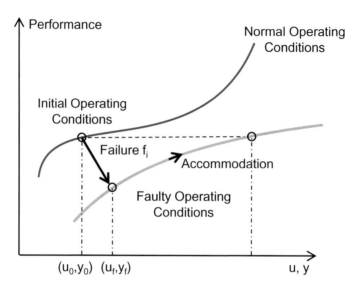

Fig. 2.4 Active fault-tolerant control

the new controller to be able to achieve a tolerable degradation of the performance in the faulty process [8]. Figure 2.4 illustrates the basic idea. After the occurrence of the fault, the system deviates from the nominal operating point of this system, which is defined by the system's input/output variables (u_0, y_0), to a faulty operating point (u_f, y_f). The objective of active fault-tolerant control is to develop an alternative control law or an alternative control structure which considers the degraded system parameters and realises a new operation point (u_0, y_0) for the system in order to preserve the main performance aspects (stability, accuracy, . . .), i.e. these aspects are ideally very close to the initial system parameters [48].

In general, systems consist of actuators, the process and sensors. Based on this distinction three classes of faults can be distinguished: actuator faults, process faults and sensor faults.

Actuator faults can be seen as any kind of malfunction of the equipment that actuates the system, such as faults of valves, pumps or electrical motors. The cause for this kind of malfunction may be abnormal operation or material degradation [62]. An actuator with additive and/or multiplicative faults can be described using the following equation:

$$u_i^{*,f} = \alpha_k^j u_j^* + u_{j0}^*. \tag{2.5}$$

In this equation, u_j^* and $u_i^{*,f}$ denote the jth normal and faulty control actions and u_{j0}^* represents a constant offset in the case that the respective actuator is blocked. α_k represents a gain degradation of the jth component. For this gain degradation, $\alpha_k = 1$ indicates a fully functional actuator and $\alpha_k = 0$ an actuator with no effect at all.

If actuator faults are present, these faulty actuators will corrupt the closed-loop behaviour. Additionally, the controller tries to cancel the error between the measurement and its reference input, which is based on fault-free conditions [62]. In these conditions, the controller gain is not "optimal" and can drive the system to its physical borders or even into instability.

Process faults occur in the case that some changes in the system render the dynamic relation invalid, e.g. leaks in a tank system or a loss of air pressure in the tire of a vehicle. In this case altered functions will describe the system behaviour. In the continuous-time domain this may be:

$$\dot{x}(t) = f_f(x(t), u(t)), \tag{2.6}$$

$$y(t) = h_f(x(t), u(t)), \tag{2.7}$$

or in the discrete-time domain:

$$x_{k+1} = f_f(x_k, u_k), \tag{2.8}$$

$$y_k = h_f(x_k, u_k). \tag{2.9}$$

In these equations $x \in \mathbb{R}^n$ is the state vector, $y \in \mathbb{R}^m$ is the output vector and $u \in \mathbb{R}^r$ denotes the control input vector. f_f and h_f are (in general) non-linear functions describing the altered behaviour in the faulty case. It is important to note that these functions will not be known in most cases and cannot easily be formulated in analytical form. For certain less-complex systems and frequent faults it is possible to formulate these functions analytically. Additionally, it is a problematic fact that the functions can change over time even rapidly when the fault gets more severe.

Sensor faults can be seen as a serious amount of variations of one or more measurements. Two main sensor fault scenarios can be distinguished: "sensor-lock-in-place" and "loss of measurement accuracy".

The notion "sensor-lock-in-place" is applicable in situations within which a sensor will be held in a certain position at an unknown time t_f and the sensor will not provide the actual and current value of the variable, which is to be measured [69]:

$$y_{i,k} = y_{i,t_f} = \text{const}, \quad \forall k > t_f. \tag{2.10}$$

This may be the case, if a longitudinal sensor is jammed and cannot follow a movement in the process any more.

The second notion "loss of measurement accuracy" is applicable if a degradation of the measurement accuracy of the sensor is present:

$$y_{i,k} = k_i y_{i,k}^c, \quad \forall k > t_f. \tag{2.11}$$

In this equation, $y_{i,k}^c$ denotes the true value of the variable to be measured and k_i is significantly different from 0 [69]. This may be the case if e.g. if the connection of a electrical sensor is corroded and leads to unwanted resistance. In some applications,

Fig. 2.5 Fault types
(behaviour over time)

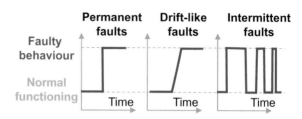

such as automotive applications, sensors which are connected by means of a bus
system (CAN bus, flexray, ethernet, ...) dispose of an integrated diagnosis system
which can also indicate sensor faults.

Faults can be also differentiated on the basis of their behaviour over time. *Permanent faults* can be observed in the form of abrupt changes, manifesting a permanent
character changing either physical parameters or system structures [65]. *Drift-like
faults* often have a long evolution over time and can be observed in the form of
slowly evolving faults. *Intermittent faults* are malfunctions with a short duration.
However this malfunctions can still lead to long-lasting effects, e.g. actuator saturations because of excessive loads. Figure 2.5 gives an overview of the different fault
types.

A general scheme of modern control systems which offer the possibilities to
accommodate the different kinds of faults is presented in Fig. 2.6 [69].

The main part of the scheme is the controlled system with its actuators, process dynamics and sensors. All parts are affected by the so-called unknown inputs.
Unknown inputs can process noise and measurement noise as well as external disturbances acting on the system. For active fault-tolerant control model-based methods
and analytical redundancy-based fault diagnosis are utilised [8, 16, 28, 34, 49, 69].

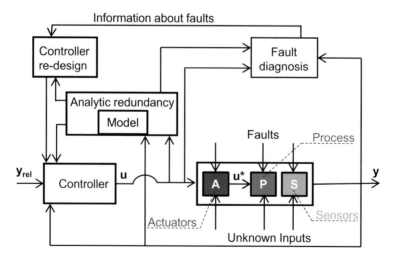

Fig. 2.6 Modern control system

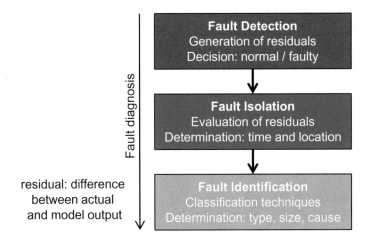

Fig. 2.7 Sub-tasks of fault diagnosis

In this case, the observed mismatch between the mathematical model and the original system under consideration can be used for fault diagnosis purposes. The mismatch can be utilised for determining that faults are present, for determining where faults are located and for determining the type of the fault and its size and cause, i.e. fault detection, fault isolation and fault identification.

The first subtask of fault diagnosis is *fault detection*. This subtask consists of the generation of signals called residuals that reflect the consistency between actual data (sensor readings from the controlled system) and the respective model (analytic redundancy). Typically residuals are defined as a difference between the outputs of the systems under consideration and their estimate obtained with the mathematical model [69]. These residuals allow to decide whether the system works under normal conditions of if one or multiple faults have occurred.

The residuals can be evaluated in the second subtask of fault diagnosis – the *fault isolation*. The residuals are logically analysed in order to find indications of the time of the occurrence of the fault(s) and the location of the fault, i.e. which actuator or sensor is faulty, or which characteristic of the process has changed because of a fault.

In the third subtask *fault identification* determines the type of fault, its size and cause. Frequently classification techniques are applied which can be based on artificial intelligence (AI) techniques such as dynamic neural networks or deep learning [29]. Figure 2.7 (compare [69]) summarises the sub-tasks of fault diagnosis.

2.1 FTC for Continuous Processes

Continuous Event Systems (CES) are systems with a continuous state that changes continuously over time, e.g. the amount of liquid in a tank and or the temperature of an electrical motor. Such systems can be described by the differential equations (2.1) and (2.2).

Fig. 2.8 Nominal and faulty
system behaviour B

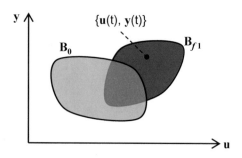

It is possible to model such systems using the *system behaviour B*, which is the congregation of all input/output (I/O) pairs, which are allowed by the respective system. For static single-input-single-output (SISO) systems, the behaviour B can be represented by curves in the I/O space. For multi-input-multi-output (MIMO) systems, the behaviour can be a map defined by the congregation of vector pairs $\{u, y\}$. Because this kind of representation is not feasible for dynamical systems, a set representation can be chosen in the congregation of the vector function pairs $\{u(t), y(t)\}$ [64]. The behaviour of the system B_{fi} in the case of a fault f_i is, in general, different from the behaviour B_0 in the nominal case. Figure 2.8 presents the nominal and faulty systems behaviour.

An analytical model can be derived by describing the system behaviour by a set of variables $z_j \in Z$ and a set of constraints $c_i \in C$ which relates these variables to each other. In the case of dynamical systems, the constraints c_i may include the relation between variables z and their time derivative $\dot{z}$ [64]. A faulty system can be described with four types of variables: unknown states $x \in X$, inputs $u \in U$, outputs $y \in Y$ and faults $f \in F$:

$$Z = X \cup U \cup Y \cup F, \tag{2.12}$$

The known variables can be described as:

$$K = U \cup Y, \tag{2.13}$$

Constraints can be described in the subsequent form:

$$c_i : 0 = h_i(x, u, y, f). \tag{2.14}$$

In this equation, h_i is a scalar function.
The mapping

$$var : 2^C \rightarrow 2^Z, \tag{2.15}$$

associates a set of constraints with the set of variables occurring in these constraints.

An operating point can be described by a tuple $(x \in X, u \in U, y \in Y)$ which does not contradict the set of constraints C. A set of operating points can be called the "operating region". The set C of constraints incorporates a qualitative fault model, with which f_i signifies a fault value. By convention, $f_i = 0 \, \forall i$ is valid in the nominal case.

Another modelling possibility for continuous time systems is a structural model or *structure graph*. The structure graph is a qualitative representation of the physical couplings between the inputs, the internal variables (e.g. state variables) and the outputs of a dynamical system [8, 64]. The structure graph is a bipartite graph

$$G = (Z \cup C, E), \tag{2.16}$$

which has two kinds of vertices which can either represent the variables (Z) or the constraints (C). In the case that the variable z_j is present in the constraint c_i, an undirected edge $e \in E$ between the vertex c_i and the vertex z_j is created in this graph. It is possible to represent the graph with its incidence matrix M. In this matrix the ijth element $m_i j$ is equal to "1" in the case that an edge between c_i and z_j is present. This can be described by:

$$m_{ij} = \begin{cases} 1, & z_j \in var(\{c_i\}) \\ 0, & z_j \notin var(\{c_i\}) \end{cases} \tag{2.17}$$

For this kind of system two different kinds of diagnosis are possible. Based on the system behaviour the *consistency* can be checked. Based on the knowledge that a set C of constraints is the congregation of the equations of the physical model of the system it can be concluded that a system is faulty, if a measured I/O pair $\{u(t), y(t)\}$ does not belong to the behaviour B_0, i.e. it contradicts the model C. However, if the measured I/O pair contradicts the model C_{fi}, which could be determined for a certain fault f_i, and only single faults are assumed, the system cannot be subject to the fault f_i (fault identification), i.e. this fault cannot be present in the system.

The second kind of diagnosis relies on an structural analysis of the structure graph. One prominent approach relies on the algorithm for finding over-constrained subsystems developed by Krysander et al. [35]. They define that a set M of equations is structurally overdetermined, if M has more equations than unknown variables and also define that a structurally overdetermined set is a "Minimal Structurally Overdetermined" (MSO) set, if no proper subset is a structurally overdetermined set. The resulting MSOs can for instance be used to identify faults which are structurally non-detectable and faults which are not structurally isolable.

In general, continuous processes are represented by continuous variables on a continuous time basis (i.e. mechanical, electrical, electro-magnetic, hydraulic physical systems, belong to this category). Several approaches for the modelling and simulation of continuous processes can be found in literature (compare e.g. [6]). The analysis of these complex processes was traditionally approached with different mathematical methods, including "Differential Algebraic Equations" (DAEs), "Ordinary Differential Equations" (ODEs), or "Partial Differential Equations" (PDEs).

However, for most complex systems the solutions to these equations are very difficult or impossible to find [66]. It is therefore often advisable to employ the elaborate approaches developed for discrete event systems (compare Sect. 2.2).

2.2   FTC for Discrete Event Systems

A "Discrete Event System" (DES) is a dynamical system driven by the instantaneous (sometimes even abrupt) occurrence of events. The state transitions of a DES indicate the physical phenomenon that causes the change in state. In general, discrete event simulation is appropriate for systems whose state is discrete and changes at particular points in time and then remains in that state for a period of time. Discrete event system approaches have been recognised as a promising framework due to the significance of event-driven models in large and complex systems [23]. The elaborate theory allows a systematic design of diagnostic systems and the computational efficiency enables on-line diagnosis even for complex systems.

The number of reported approaches for modelling and simulating DESs is ever growing. Prominent approaches include the application of robust H∞-Based Approaches [69], of fuzzy logic [12, 42, 69], of probabilistic finite automata [33], of petri nets [21], of multi-agent systems [76] or the distinction between intermittent and permanent faults [23]. Additionally systems for the creation, verification and synthesis of DESs are presented such as *Supremica* [47] or *UltraDES* [2]. In recent years also a combination of continuous and discrete-event simulation has been intensively researched under the notion "hybrid systems" [22]. Several research initiatives concern the fault diagnosis of DESs, which can be classified into Petri net based methods and automata based methods [20]. A current overview of Petri net based methods is given by Giua and Silva [21]. For Petri net based methods integer linear programming [15] and partially observed Petri nets have been applied in order to avoid complexity [39]. Computational advantages for automaton based methods have been recently reported based on N-diagnosability [41].

An effective and time-scheduled performance of a dynamical system requires an appropriate modelling framework. In this project, the complex system is perceived as in Cassandras et al. [11], who introduce a "Discrete Event System" (DES) as a special class of systems through merging discrete-state systems and event-driven systems. An event-driven system bases upon the concept of events. These events should be thought as occurring instantaneously and causing transitions from one state value to another. The state transitions of a DES indicate the physical phenomenon which caused the respective change in the state. For instance, in a communication protocol typical event labels are "time out", "packet received", "packet sent", while in a manufacturing system, events of interest are "machine breakdown", "machine repaired", "part accepted", etc. There are many areas in which DESs arise and the different aspects of behavior relevant in each area have led to the development of a variety of DES models. Indeed, DES can be modelled by "Finite State Automata" (FSA), extended state machines, Petri nets, event-graphs, formal languages, generalised semi

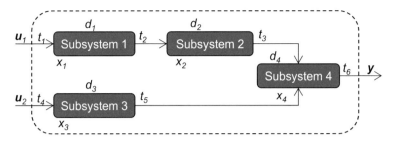

Fig. 2.9 An exemplary complex system

Markov processes as well as non-linear programming [1, 26, 43, 54, 70]. In order to restrict the behavior of a DES to a desired scope, supervisory control theory is required [53] which has been already investigated within several contexts [5, 14, 50, 63]. It is important to mention that the representation of DES may exclusively process classical state transitions (i.e. without uncertainty). However, in practice, it is rather easy to identify situations within which the state transitions of systems can be imprecise, uncertain and even unclear. If one wants to solve such situations, for example the recently developed extension of classical DES to fuzzy DES can be applied [37, 43]. It is important to mention that such systems have been successfully applied in communication systems, networked systems, manufacturing systems and automated traffic systems [36, 55, 61]. It should also be emphasised that models which describe a DES are always non-linear in conventional algebra. However, one may define a class of discrete-event systems, usually called the max-plus linear discrete-event systems, in which a synchronization without concurrency or selection is present. This class of DES can be represented by linear models using the max-plus algebra [4, 10]. This fact justifies its application, instead of more traditional approaches, for modelling complex systems [68] (exemplified in Fig. 2.9).

The $(max, +)$ algebraic structure $(\mathbb{R}_{max}, \oplus, \otimes)$ can be defined using the following equations: $\mathbb{R}max = \mathbb{R} \cup \{-\infty\}$,

$$\forall_{a,b\in\mathbb{R}_{max}} a \oplus b = max(a, b),$$
$$\forall_{a,b\in\mathbb{R}_{max}} a \otimes b = a + b,$$
$$\forall_{a,b\in\mathbb{R}_{max}} a \otimes (-\infty) = (-\infty) \otimes a = (-\infty).$$

In these equations, $\mathbb{R}$ denotes the field of real numbers. The above operators (along with their vector/matrix extension) allow describing a complex system in the state-space form:

$$x(k + 1) = A \otimes x(k) \oplus B \otimes u(k) \tag{2.18}$$
$$y(k) = C \otimes x(k). \tag{2.19}$$

In these equations, the index k is the event counter, while:

- $x(k) \in \mathbb{R}_{\text{max}}^n$ is a state vector including the time instants at which the internal events occur for the kth event counter,

- $u(k) \in \mathbb{R}^r_{\max}$ is a vector consisting of time instants at which the input events occur for the kth event counter,
- $y(k) \in \mathbb{R}^m_{\max}$ is a vector consisting of time instants at which the output events occur for the kth event counter,

and the system matrices are $A \in \mathbb{R}^{n \times n}_{\max}$, $B \in \mathbb{R}^{n \times r}_{\max}$, and $C \in \mathbb{R}^{m \times n}_{\max}$. Another superiority of max-plus algebra framework over the most popular modelling frameworks like, e.g. Petri nets, is the fact that it can be used for analysing the complex system fundamental properties, predict its behavior in the future and on-line synthesise its control strategy. This makes it possible to design effective control strategies such as "Model Predictive Control" (MPC) for max-plus systems [13], which takes into account system performance indices and constraints related to all its relevant variables. Moreover, the above MPC framework can also be extended to cope with parameter uncertainties of the matrices A, B and C [46, 56]. Having a description of the complex system, it is possible to go one level down into the subsystems. An early detection of subsystem faults as well as their proper maintenance can help to avoid system shut-down, breakdowns and even catastrophes involving human fatalities and material damage. Thus, subsystem actuator and sensor fault estimation problems along with FTC design constitute important parts of the complex system framework. Nevertheless, FTC cannot alter the reliability of the components of a plant, but it has the capability to increase the overall reliability, because FTC enables the subsystem to remain functional even after the faults have occurred.

In the field of assembly processes, the objective of such complex systems can be to process pre-products in order to achieve the final product. An exemplary complex system is portrayed in Fig. 2.9.

It is important to underline the fact that the ith subsystem should start its operation at time x_i and should accomplish its mission within the time d_i. The accomplishment of this kind of mission is a necessary condition for starting a mission of a subsequent subsystem (compare subsystem 1 and subsystem 2). This iterative behaviour is performed according to a predefined schedule x_i^r describing the desired evolution of all x_i. One should additionally note that u_i is the time of providing appropriate pre-products to realise the mission of the ith inlet subsystem (compare subsystem 1 and subsystem 3). In this context, t_i is the time of transferring the preprocessed products among the subsystems. Finally, y stands for the time at which the final product is available. This typical situation can be observed in various industries. Moreover, each subsystem is composed of actuators, sensors and the process itself representing the functional objective, while v, z, s represent its input, state and output, respectively (compare Fig. 2.10).

Thus, the performance of these components is crucial for the entire complex system. A significant research effort has been focused on the design of fault diagnosis

Fig. 2.10 Subsystem structure

of sensors and actuators with computer-based counterparts. However, the design of such schemes has only been oriented towards compensating the effect of sensor and actuator malfunctions by performing appropriate control actions called Fault-Tolerant Control (FTC) [8, 30, 49, 69]. Such strategies can be perceived as post-fault ones. Current approaches intend to create pro-active strategies (compare Sect. 2.5).

2.3 Fault Identification—A Fundamental Tool for Active FTC

The main task of *fault identification* is to determine the type of fault as well as the size and the cause of the fault. The faulty system can be represented by the Eqs. (2.20) and (2.21):

$$x_{f,k+1} = Ax_{f,k} + \sum_{i=1}^{s} A_{f,i} f_{P,i,k} x_{f,k} + Bu_{f,k} + L_1 f_{A,k}, \qquad (2.20)$$

$$y_{f,k+1} = Cx_{f,k+1} + L_2 f_{S,k}. \qquad (2.21)$$

The elements of the Eqs. (2.20) and (2.21) are: $x_{f,k} \in \mathbb{R}^n$ is the state of the respective system, $y_{f,k} \in \mathbb{R}^m$ stands for the output of this system, $u_{f,k} \in \mathbb{R}^r$ denotes the input of this system, $A_{f,i}$ denotes the distribution matrix of the ith process fault $f_{P,i} \in \mathbb{R}_p^n$ where n_p is the number of process faults, $f_{A,k} \in \mathbb{R}^s$, ($s \leq m$) is the actuator fault vector and L_1 represents the distribution matrix of the ith actuator fault, which can also be assumed to be known. $f_{S,k} \in \mathbb{R}^t$, ($t \leq m$) is the sensor fault vector and L_2 represents the distribution matrix of the ith sensor fault, which can again be assumed to be known.

A straight-forward approach for fault identification for actuator faults is presented by Dziekan et al. [17]. They are able to use estimates of the faulty state $\hat{x}_{f,k}$ in order to achieve a actuator fault estimate $\hat{f}_{A,k}$ and an associated fault estimation error. A method which allows less restrictive assumptions concerning the properties of the disturbances influencing the diagnosed system and uncertainty of its model is proposed by Buciakowski et al. [9]. For process faults, Pazera and Witczak [51] propose a process fault estimator (PFE), which is based on the quadratic boundedness approach and enables simultaneous state and process fault estimation. For the estimation of sensor faults an approach proposed by Pazera et al. [52] can be applied, which simultaneously estimates actuator and sensor faults and relies on a Takagi–Sugeno-based $H\infty$ fault estimator.

Frequently, classification techniques are applied which can be based on "Artificial Intelligence" (AI) techniques such as dynamic neural networks or deep learning [29]. Other approaches use time-frequency methods in connection with feature extraction [3], the fault characteristic "Product Function Correntropy" (PFC) in connection with the application of a "Least Square Support Vector Machine" (LSSVM) [19] and the D-S evidence theory [72].

2.4 Fault-Tolerant Controllers

When investigating fault-tolerant controllers, one needs to distinguish the case of passive or active FTC (compare Fig. 2.3). The design of passive fault-tolerant controllers assures that the closed-loop system satisfies given performance requirements for all admissible uncertainties and faults. Active fault-tolerant controllers use the results of FDI and change either parameters or the configuration of the controller in the case of faults. Prominent research approaches concern reconfiguration schemes [8, 69]. Fault-tolerance is achieved by system and/or controller reconfiguration, in which the overall system performance will be recovered (possibly to an acceptable degree) by a reconfiguration of parts of the control system under real-time condition [44, 73], after faults are identified and a reduction of the system performance is observed [8, 16]. This is a new challenge in the field of control engineering. In the last years, a considerable number of FTC approaches have been reported and a majority of them have been developed for some special applications and the resulting FTC systems are often a bank of controllers supported by an FDI system [18]. The structure of a logic-based switching controller with a bank of controllers is shown in Fig. 2.11.

Over the last years, a number of research approaches concern active fault tolerant controllers. The research of Yang et al. [71] focuses on the design of a nominal controller and a fault accommodating controller based on the synthesis of an H∞ robust controller. In this case, it is assumed that faults are uncertainties introduced as additive disturbance to the sensor inputs.

The development of fault-tolerant controllers for multi-agent systems is reported by Khalili et al. [31]. In their development, each agent disposes of a local FTC scheme which includes a fault diagnosis module and a reconfigurable controller module, consisting of a baseline controller and also two adaptive fault-tolerant controllers,

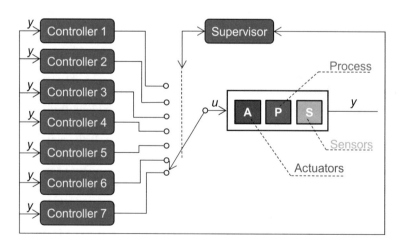

Fig. 2.11 Structure of a logic based switching controller

which are active in case of faults. Han et al. [24] propose a fault-tolerant controller, based on a a fuzzy reduced-order state/fault estimation observer to estimate the system state, sensor, process and actuator faults and which can compensate the effect of faults and guarantee the robust stability of the closed-loop system.

In spite of these current research activities, a distinct lack of a framework of FTC technology can still be observed, based on which a systematic design of FTC systems could be carried out.

2.5 Prognosis of Faults

The estimation of the "Remaining Useful Life" (RUL—compare Sect. 1.2) can be used for predicting the "Mean-Time-To-Failure" (MTTF) of the system components as well as for controlling and maintaining the system in a manner which allows to enhance the MTTF. Extensive research was carried out in order to develop algorithms for the estimation of the RUL; for surveys see [25, 32, 38, 40, 57]. Figure 2.12 presents a typical system component degradation behaviour along with the RUL-associated degradation model.

Thus, the RUL of a subsystem is defined as the length from the current point in time to the end of the useful life, i.e. the point of crossing the failure threshold. The phenomenon of degradation and ageing mechanisms have been intensively researched in the last years [7, 59]. Several sources of information are necessary in order to determine the RUL of a subsystem systems. Based on the considerations of Sikorska et al. [58], the following questions can be formulated:

- Is a subsystem in a degraded state?
- Which condition has initiated the degradation?
- How severe is the degradation—is the subsystem on a particular degradation curve?
- How quickly is the degradation of the subsystem expected to progress from its current state to functional failure?

Fig. 2.12 Degradation modelling

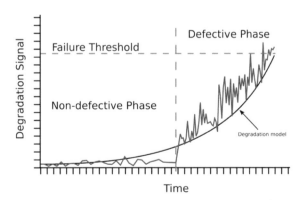

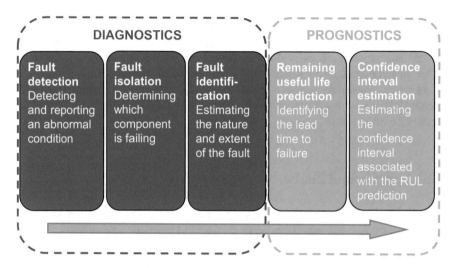

Fig. 2.13 Diagnostics-prognostics steps

- What novel events will decelerate or accelerate the expected degradation behaviour?
- How may other factors affect the estimate of the RUL?

A systematic approach towards RUL estimation adds two steps to the well-known diagnostics approach (Fig. 2.13).

When looking at the fourth stage (remaining useful life prediction) four steps can be distinguished (compare Fig. 2.14, which were identified by Lei et al. [40] in the scope of a large scale review of machinery health prognostics).

The four steps or technical processes are data acquisition, "Health Indicator" (HI) construction, "Health Stage" (HS) division and remaining useful life (RUL) prediction. Initially, measured data (e.g. vibration signals) are acquired from sensors in order to monitor the health condition of machinery. Secondly, HIs are constructed using signal processing techniques or artificial intelligent (AI) techniques in order to represent the health condition of the machinery. Thirdly, on the basis of the varying degradation trends of HIs, the whole lifetime of machinery is divided into two or more different HSs. In the HS, which presents the obvious degradation trend, it is finally possible to predict the RUL employing the analysis of degradation trends and a pre-specified "Failure Threshold" (FT).

The first step—data acquisition—concerns the processes of capturing data (measurements) from various sensors installed in the respective system of interest. Usually, various (sometime redundant) sensors are applied in order to capture the measurements necessary to reflect the current state of degradation of the system. In spite of the necessity of this step, it was not yet the point of main emphasis of research (compare [40]). Also the question, how a system should be designed in order to allow the placement of the right sensors at the best position and to ease the capturing of measurement has not yet been discussed in detail.

Fig. 2.14 Four steps for RUL prediction

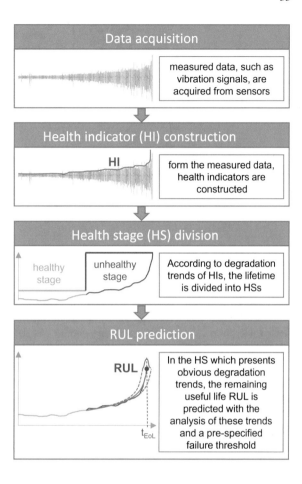

For the second step—HI construction—two strategies can be distinguished: physics based HIs which are directly related to the physical phenomena underlying the approaching failure and virtual HIs which are result from sensor fusion and have lost the direct connection to these physical phenomena.

The objective of the third step—HS division—is to divide the more or less continuous degradation process of a system into distinct stages which display a certain trend of the HIs. In some cases up to three stages are necessary for a sensible depiction of the complete degradation behaviour of a system (healthy stage; degradation stage; critical stage); this can be the case, if certain self-healing mechanisms occur and lead to an "increase-decrease-increase" degradation (compare [67]).

The fourth stage—RUL prediction—tries to predict the amount of time until the system will cross a failure threshold and has been, since many years, in the centre of numerous research activities. However, approaches to increase the RUL of complex systems which consists of several subsystems has not yet been in the focus of these research activities.

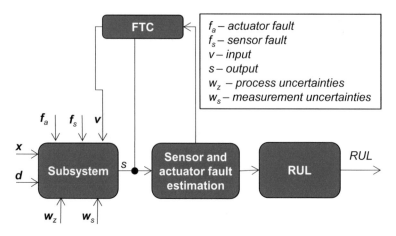

Fig. 2.15 Low-level layer of the ith subsystem

It can be a general research objective to develop cooperative strategies increasing the RUL of complex systems by optimised exploitation and monitoring of all their subsystems. A two-layer strategy can be adopted to tackle such a challenging objective. The structure of the low-level layer is presented in Fig. 2.15).

It can be observed that the ith subsystem is fed with a scheduled starting time x_i, and d_i. As a result of realizing a desired action, actual performance parameters are measured, i.e. x_i^m and d_i^m. It is important to note that they are usually different from the scheduled ones. The size of this discrepancy depends on the health-oriented efficiency of subsystem components, i.e. actuators and sensors. Thus, the crucial part of the entire low-level layer is the sensor and actuator fault estimation strategy, which provides interval fault estimates describing the estimates along with their uncertainty [69]. Based on fault estimates, a suitable FTC action is performed in order to avoid a failure of ith subsystem. While sensor measurements of s can be (up to some extent) replaced by their estimates obtained with, e.g. observers, actuators cannot be replaced in the same fashion. Thus, the actuator fault estimate is used as a degradation signal (compare Fig. 2.13) to develop a degradation model. Subsequently, this model is used for predicting the RUL of the ith subsystem. The structure of the top-level layer is presented in Fig. 2.16.

The objective of the feasible schedule calculator is to obtain the fastest possible schedule x^f based on the complex system behaviour expressed in actual performance parameters being measured. This information along with the RUL vector of all subsystems is provided to the user interface. Taking into account the fact that the more demanding schedule of a given subsystem is, the lower is its RUL (premature exploitation) and the user may slow down its fastest possible schedule. As a result, a new schedule x^r is provided to the complex system controller, which calculates appropriate pre-products providing times u_i. It is therefore possible to choose a schedule consciously, which can increase the RUL of the subsystems.

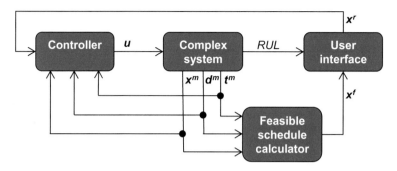

Fig. 2.16 Top-level layer

2.6 Summary

In this section the main aspects and components of fault-tolerant control have been explained. It is important to note that, by an addition of two prognostics steps, it is possible to be aware about upcoming faults and that scheduling strategies are possible which will prolong the RUL of complex systems.

References

1. Abrams, M., Doraswamy, N., Mathur, A.: Visual analysis of parallel and distributed programs in the time, event, and frequency domains. IEEE Trans. Parallel Distrib. Syst. **3**(3), 672–685 (1992)
2. Alves, L.V.R., Martins, L.R.R., Pena, P.N.: Ultrades - a library for modeling, analysis and control of discrete event systems. IFAC PapersOnLine **50**(1), 5831–5836 (2017)
3. Attoui, I., Fergani, N., Boutasseta, N., Oudjani, B., Deliou, A.: Structural reliability analysis with imprecise random and interval fields. J. Sound Vib. **397**, 241–265 (2017)
4. Baccelli, F., Cohen, G., Olsder, G.J., Quadrat, J.P.: Synchronization and linearity: an algebra for discrete event systems. J. Oper. Res. Soc. **45**, 118–119 (1994)
5. Balemi, S.: Input/output discrete event processes and communication delays. Discret. Event Dyn. Syst. **4**(1), 41–85 (1994)
6. Bandyopadhyay, S., Bhattacharya, R.: Discrete and Continuous Simulation: Theory and Practice. CRC Press, Boca Raton (2017)
7. Barre, A., Deguilhem, B., Grolleau, S., Gerad, M., Suard, F., Riu, D.: A review on lithium-ion battery ageing mechanisms and estimations for automotive applications. J. Power Sources **241**, 680–689 (2013)
8. Blanke, M., Kinnaert, M., Lunze, J., Staroswiecki, M.: Diagn. Fault-Toler. Control. Springer, New York (2016)
9. Buciakowski, M., Witczak, M., Mrugalski, M., Theilliol, D.: A quadratic boundedness approach to robust dc motor fault estimation. Control Eng. Pract. **66**, 181–194 (2017)
10. Butkovic, P.: Max-Linear Systems: Theory and Algorithms. Springer, Berlin (2010)
11. Cassandras, C.G., Lafortune, S.: Introduction to Discrete Event Systems. Springer, Berlin (2008)
12. Chen, X., Xing, H.: Nonblocking check in fuzzy discrete event systems based on observation equivalence. Fuzzy Sets Syst. **269**, 47–64 (2015)

13. de Schutter, T., van den Boom, T.: Model predictive control for max-plus-linear discrete event systems. Automatica **37**(7), 1049–1056 (2001)
14. Debouk, R., Lafortune, S., Teneketzis, D.: On the effect of communication delays in failure diagnosis of decentralized discrete event systems. Discret. Event Dyn. Syst. Theory Appl. **13**(3), 263–289 (2003)
15. Dotoli, M., Fanti, M., Mangini, A., Ukovich, W.: On-line fault detection in discrete event systems by Petri nets and integer linear programming. Automatica **45**(11), 2665–2672 (2009)
16. Ducard, G.: Fault-tolerant Flight Control and Guidance Systems: Practical Methods for Small Unmanned Aerial Vehicles. Springer, Berlin (2009)
17. Dziekan, L., Witczak, M., Korbicz, J.: Active fault-tolerant control design for Takagi-Sugeno fuzzy systems. Bull. Pol. Acad. Sci. Tech. Sci. **59**(1), 93–102 (2011)
18. Farias de Santos, C.H., Cardozo, D.I.K., Polycarpou, M., Parisini, T., Cao, Y.: Bank of controllers and virtual thrusters for fault-tolerant control of autonomous underwater vehicles. Ocean Eng. **121**, 210–223 (2016)
19. Fu, Y., Jia, L., Qin, Y., Yang, J.: Product function correntropy and its application in rolling bearing fault identification. Measurements **97**, 88–99 (2017)
20. Gao, Z., Cecati, C., Ding, S.X.: A survey of fault diagnosis and fault-tolerant techniques part I: fault diagnosis with model-based and signal-based approaches. IEEE Trans. Ind. Electron. **62**(6), 3757–3767 (2015)
21. Giua, A., Silva, M.: Modeling, analysis and control of discrete event systems: a petri net prespective. IFAC PapersOnLine **50**(1), 1772–1783 (2017)
22. Grastien, A., Trave-Massuyes, L., Puig, V.: Solving diagnosability of hybrid systems via abstraction and discrete event techniques. IFAC PapersOnLine **50**(1), 5023–5028 (2017)
23. Guanqian, D., Jing, Q., Guanjun, L., Kehong, L.: A discrete event systems approach to discriminating intermittent from permanent faults. Chin. J. Aeronaut. **27**(2), 390–396 (2014)
24. Han, J., Zhang, H., Wang, Y., Liu, X.: Robust state/fault estimation and fault tolerant control for ts fuzzy systems with sensor and actuator faults. J. Frankl. Inst. **353**, 615–641 (2016)
25. Heng, A., Zhang, S., Tan, A.C.C., Mathew, J.: Rotatin machinery prognostics: state of the art, challenges and opportunities. Mech. Syst. Signal Process. **23**, 724–739 (2009)
26. Hillion, H.P., Proth, J.M.: Performance evaluation of job-shop systems using timed eventgraphs. IEEE Trans. Autom. Control **34**(1), 3–9 (1989)
27. Isermann, R.: Fault Diagnosis Systems. An Introduction from Fault Detection to Fault Tolerance. Springer, New York (2006)
28. Isermann, R.: Fault Diagnosis Applications: Model Based Condition Monitoring, Actuators, Drives, Machinery, Plants, Sensors, and Fault-Tolerant Systems. Springer, Berlin (2011)
29. Jiang, Y., Qinglei, H., Ma, G.: Gearbox fault identification and classification with convolutional neural network. Shock Vib. **2015**(390134), 1–10 (2015)
30. Johansen, T.A., Fossen, T.I.: Control allocation a survey. Automatica **49**(5), 1087–1103 (2013)
31. Kahlili, N., Zhang, X., Polycarpou, M., Parisini, T., Cao, Y.: Distributed adaptive fault-tolerant control of uncertain multi-agent systems. IFAC-PapersOnLine **48–21**, 66–71 (2015)
32. Kan, M.S., Tan, A.C.C., Mathew, J.: A review on prognostic techniques for non-stationary and non-linear rotating systems. Mech. Syst. Signal Process. **62–63**, 1–20 (2015)
33. Keroglu, C., Hadjicostis, C.: Detectability in stochastic discrete event systems. In: 12th IFAC/IEEE Workshop on Discrete Event Systems, pp. 27–32 (2014)
34. Korbicz, J., Kościelny, J., Kowalczuk, Z., Cholewa, W. (eds.): Fault Diagnosis. Models, Artificial Intelligence, Applications. Springer, Berlin (2004)
35. Krysander, M., Aslund, J., Nyberg, M.: An efficient algorithm for finding minimal overconstrained subsystems for model-based diagnosis. IEEE Trans. Syst. Man Cybern. **38**(1), 197–206 (2008)
36. Kumar, R., Garg, V.K.: Modeling and Control of Logical Discrete Event Systems, vol. 300. Springer Science and Business Media, Berlin (2012)
37. Lamperti, G., Zanella, M.: Flexible diagnosis of discrete-event systems by similarity-based reasoning techniques. Artif. Intell. **170**(3), 232–297 (2006)

38. Lee, J., Wu, F., Zhao, W., Ghaffari, M., Liao, L., Siegel, D.: Prognostics and health management design for rotary machinery systems - reviews, methodology and applications. Mech. Syst. Signal Process. **42**, 314–334 (2014)
39. Lefebvre, D.: On-line fault diagnosis with partially observed petri nets. IEEE Trans. Autom. Control **59**(7), 1919–1924 (2015)
40. Lei, Y., Li, N., Guo, L., Li, N., Yan, T., Lin, J.: Machinery health prognostics: a systematic review from data acquisition to RUL prediction. Mech. Syst. Signal Process. **104**, 799–834 (2018)
41. Lin, F., Wang, L.Y., Chen, W., Han, L., Shen, B.: N-diagnosability for active on-line diagnosis in discrete event systems. Automatica **83**, 220–225 (2017)
42. Liu, F., Dziong, Z.: Decentralized diagnosis of fuzzy discrete-event systems. Eur. J. Control **3**, 304–315 (2012)
43. Liu, J., Li, Y.: The relationship of controllability between classical and fuzzy discrete-event systems. Inf. Sci. **178**(21), 4142–4151 (2008)
44. Mahmoud, M., Jiang, J., Zhang, Y.: Active Fault Tolerant Control Systems: Stochastic Analysis and Synthesis. Springer, Berlin (2003)
45. Mahulkar, V.V.: Structural technology evaluation and analysis program (STEAP) Delivery order 0037: prognosis-based control reconfiguration for an aircraft with faulty actuator to enable performance in a degraded state. United States Air Force, 2010
46. Majdzik, P., Akielaszek-Witczak, A., Seybold, L., Stetter, R., Mrugalska, B.: A fault-tolerant approach to the control of a battery assembly system. Control Eng. Pract. **55**, 139–148 (2016)
47. Malik, R., Akesson, K., Flordal, H., Fabian, M.: Supremicaan efficient tool for large-scale discrete event systems. IFAC PapersOnLine **50**(1), 5794–5799 (2017)
48. Noura, H., Sauter, D., Hamelin, F., Theilliol, D.: Fault-tolerant control in dynamic systems. Application to a winding machine. IEEE Control Syst. Mag. **20**(1), 33–49 (2000)
49. Noura, H., Theilliol, D., Ponsart, J., Chamseddine, A.: Fault-Tolerant Control Systems: Practical Applications. Springer, Berlin (2013)
50. Park, S.J., Cho, K.H.: Delay-robust supervisory control of discrete-event systems with bounded communication delays. IEEE Trans. Autom. Control **51**(5), 911–915 (2006)
51. Pazera, M., Witczak, M.: Towards robust process fault estimation for uncertain dynamic systems. In: Proceedings of the 21st International Conference on Methods and Models in Automation and Robotics (MMAR) (2016)
52. Pazera, M., Witczak, M., Buciakowski, M., Mrugalski, M.: Simultaneous estimation of multiple actuator and sensor faults for Takagi–Sugeno fuzzy systems. In: Proceedings of the 22nd International Conference on Methods and Models in Automation and Robotics (MMAR) (2017)
53. Ramadge, P.J.G., Wonham, M.W.: Supervisory control of a class of discrete event processes. SIAM J. Control Optim. **62**(6), 206–230 (1987)
54. Sahner, R.A., Trivedi, K., Puliafito, A.: Performance and Reliability Analysis of Computer Systems: An Example-based Approach Using the SHARPE Software Package. Springer Science and Business Media, Berlin (2012)
55. Seatzu, C., Silva, M., van Schuppen, J.: Control of Discrete-Event Systems. Lecture Notes in Control and Information Sciences, vol. 433. Springer, Berlin (2012)
56. Seybold, L., Witczak, M., Majdzik, P., Stetter, R.: Towards robust predictive fault-tolerant control for a battery assembly system. Int. J. Appl. Math. Comput. Sci. **25**(4), 849–862 (2015)
57. Si, X.-S., Wang, W., Hu, C.-H., Zhou, D.-H.: Remaining useful life estimationa review on the statistical data driven approaches. Eur. J. Oper. Res. **213**(1), 1–14 (2011)
58. Sikorska, J.Z., Hodkiewicz, M., Ma, L.: Prognostic modelling options for remaining useful life estimation by industry. Mech. Syst. Signal Process. **25**, 1803–1836 (2011)
59. Stetter, R., Witczak, M.: Degradation modelling for health monitoring systems. In: Proceedings of the Conference of Advanced Control and Diagnosis ACD 2014 (2014)
60. Tabatabaeipour, S.M.: Fault Diagnosis and Fault Tolerant Control of Hybrid Systems. Aalborg University, Aalborg (2010)
61. Takai, S., Kumar, R.: Distributed failure prognosis of discrete event systems with bounded-delay communications. IEEE Trans. Autom. Control **57**(5), 1259–1265 (2012)

62. Theilliol, D., Cedric, J., Zhang, Y.: Actuator fault tolerant control design based on a reconfigurable reference input. Int. J. Appl. Math. Comput. Sci. **18**(4), 553–560 (2008)
63. Tripakis, S.: Decentralized control of discrete-event systems with bounded or unbounded delay communication. IEEE Trans. Autom. Control **49**(9), 1489–1501 (2004)
64. Ungermann, M., Lunze, J., Schwarzmann, D.: Test signal generation for service diagnosis based on local structural properties. Int. J. Appl. Math. Comput. Sci. **22**(1), 55–65 (2012)
65. Varga, A.: Solving Fault diagnosis Problems. Linear Synthesis Techniques. Springer, Berlin (2017)
66. Wainer, G.A., D'Abreu, M.C.: Using a discrete-event system specifications (DEVS) for designing a modelica compiler. Adv. Eng. Softw. **79**(1), 111–126 (2015)
67. Williams, T., Ribadeneira, X., Billington, S., Kurfess, T.: Rolling element bearing diagnostics in run-to-failure lifetime testing. Mech. Syst. Signal Process. **15**(5), 973–993 (2001)
68. Witczak, M.: Modelling and Estimation Strategies for Fault Diagnosis of Non-linear Systems. Springer, Berlin (2007)
69. Witczak, M.: Fault Diagnosis and Fault-Tolerant Control Strategies for Non-linear Systems. Analytical and Soft Computing Approaches. Springer, Berlin (2014)
70. Yan, F., Dridi, M., El Moundi, A.: An autonomous vehicle sequencing problem at intersections: a genetic algorithm approach. Int. J. Appl. Math. Comput. Sci. **23**(1), 183–200 (2013)
71. Yang, S.S., Chen, J., Mohamed, H.A.F., Moghavvemi, M.: Sensor fault tolerant controller for a double inverted pendulum system. In: Proceedings of the 17th World Congress The International Federation of Automatic Control, pp. 2588–2594 (2008)
72. Yin, N., Xing, J., Liu, Y., Li, Z., Lin, X.: A novel single-phase-to-ground fault identification and isolation strategy in wind farm collector line. Electr. Power Energy Syst. **94**, 15–26 (2018)
73. Yin, S., Luo, H., Ding, S.X.: A survey of fault-tolerant controllers based on safety-related issues. IEEE Trans. Ind. Electron. **61**(5), 2402–2411 (2014)
74. Zhang, Y., Jiang, J.: Bibliographical review on reconfigurable fault-tolerant control systems. Annu. Rev. Control **32**(2), 229–252 (2008)
75. Zhao, J., Chen, B., Shen, J.: Multidimensional nonorthogonal wavelet-sigmoid basis function neural networkfor dynamic process fault diagnosis. Comput. Chem. Eng. **23**(1), 83–92 (1998)
76. Zhu, W., Pu, H., Wang, D., Li, H.: Event-based consensus of second-order multi-agent systems with discrete time. Automatica **79**, 78–83 (2017)

Chapter 3
Fault-Tolerant Design

Fault-Tolerant Design (FTD) aims at supporting the engineering design of technical systems in order to increase their fault-tolerance. The approaches to increase fault-tolerance can target the observability and controllability of technical systems, but can also aim at intensifying their inherent fault-tolerant design qualities, for instance by means of applying robust physical effects. The importance of design for fault-tolerance is highlighted by Rouissi and Hoblos [44]. They assign the ability of a system to tolerate a fault to the quality of the design of this system.

FTD shares some objectives, strategies and methodical components with the design guidelines "Design for Reliability" (DfR) and "Design for Safety" (DfS). Their main objective to increase the reliability respective safety (compare Sect. 1.2) of technical systems is also targeted by FTD, but FTD has a distinct main focus—the increase of fault-tolerance.

FTD includes the design of fault-tolerant controllers (compare Sect. 2.4) but goes beyond by also including the design of sensors, actuators, the controlled processes and elements out-side of the control-loops, if they can contribute to an increase of fault-tolerance.

The design of technical systems takes usually place in industrial companies which develop and produce these technical systems. It is an obvious fact that today very often sub-contractors and suppliers are included in these processes. However, the main coordination and concept development is usually carried out at the original equipment manufacturer (OEM). Procedure models are applied in order to steer these design processes. In design science, several procedure models were proposed over the last decades (compare [15, 30, 37]). The fact cannot be denied that control and diagnosis are two main capabilities of fault-tolerant systems, therefore it is sensible to take procedure models into consideration which focus on these capabilities. Isermann [27] proposes a development scheme for fault detection and and fault diagnosis (Fig. 3.1) which is based on the V-model known from software engineering.

This model depicts the process from requirements over several steps such as modelling, simulation, testing, system integration and system testing towards the complete product which can be produced. During these steps the technical system

© Springer Nature Switzerland AG 2020   39
R. Stetter, *Fault-Tolerant Design and Control of Automated Vehicles and Processes*, Studies in Systems, Decision and Control 201,
https://doi.org/10.1007/978-3-030-12846-3_3

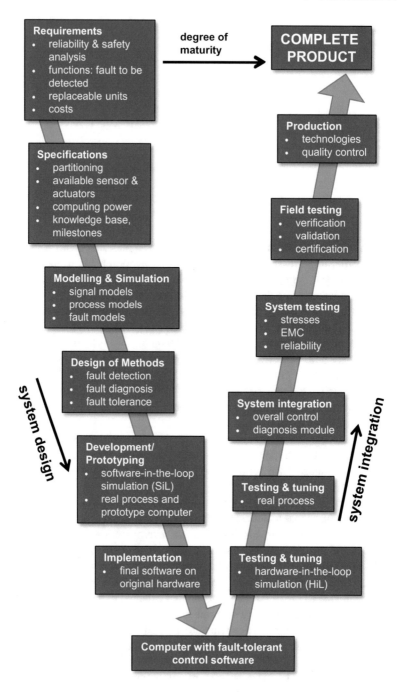

Fig. 3.1 "V" development scheme for fault detection and diagnosis systems

Fig. 3.2 Supporting processes "carry" the "V" scheme

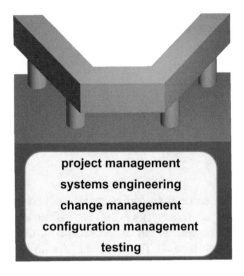

under development and its fault-tolerance capabilities become more and more mature. In the industrial companies and their sub-contractors and suppliers these processes are often obscured by further processes which are necessary in order to allow the design of technical systems and are often carried out by the same people who also have to develop the technical system. These processes "carry" the "V" development scheme (Fig. 3.2).

Five groups of activities are important in every phase of the development scheme (compare [50]). The essential activities of the "project management" focus on people, time and money and aim at planning and controlling all tasks and schedules, as well as at the assignment of resources. The logical relationships of the product and the process are dealt within "systems engineering", which can be defined as a mind-set and guidelines for the functional design of complex systems, which is based on certain thinking models and principles [12]. The notion "change management" names the processes of requesting, planning, implementing, and evaluating of changes to a system. It has two main goals: supporting the processing of changes and enabling the traceability of changes. The term "configuration management" denotes an extension of variant management [11]. Possible variants of a system have to be consciously defined so that each possible system variant (configuration) can fulfil the functional and physical requirements throughout its life. The notion "testing" summarises a number of activities which have to be planned, carried out and controlled during all phases of the core process. The activities of testing consist of virtual and physical analyses of the system under development.

In order to substantiate the discussion in this section, it was decided to use a structure concerning the level of abstraction which can be found throughout the design methodology literature [15, 30, 37]. In the last decades design science has found that the model of the level of abstraction is not appropriate as a strict time line of a design process [14], but the logical validity of this model has not been charged.

Fig. 3.3 Model of product concretization

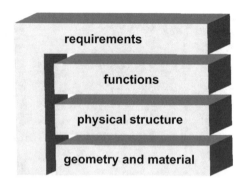

In essence, the model of the level of abstraction describes different points of view of the same technical system. The highest level of abstraction is the requirement model which describes the functional, performance and structural characteristics a product should provide. The next level "function" allows to describe on a a bit more concrete level how a technical system works, i.e. how the requirements can be achieved. On this level different kinds of function models are applied (structure-oriented, flow-oriented, or relation-oriented). The next more concrete level is represented by the "physical effects" which describe how the functions of the technical systems can be achieved in the physical domain. Even more concrete, the "geometry and the material" describe how the physical effects are realised. One prominent representation is given by the so-called Munich Model of Product Concretization (Fig. 3.3) [38].

The next sub-sections are arranged according to this model.

3.1 Requirements Exploration

Requirements are a decisive factor in all kinds of design of technical systems (compare e.g. [4]). Four out of ten top risks in system development projects are connected with requirements [24]. An early and profound exploration of requirements is one main cornerstone of successful design in leading industrial companies [4]. Requirements can be defined as the purpose, goals, constraints and criteria associated with a design project [32] and should be consciously formulated. Specific requirements for fault-tolerant design usually concern the type and amount of known possible faults towards which the technical system should be able to tolerate and the specific objective to achieve under faulty conditions certain stability and performance goals of the technical system. For the formulation of such requirements it is sensible to distinguish between possible faults and expected faults. The set of expected faults is a sub-set of the set of possible faults. The set of possible faults is usually very large even for systems with moderate complexity. Firstly, each component of a system can be faulty. Some components can even have multiple faults, e.g. a bearing can dispose of a lubrication fault which initially will only increase friction or a surface fault which will increase vibrations. Secondly, all connections between components can be faulty. Thirdly, some faults may occur because of a certain problem in the interplay between components or connections. From this large number some faults can be chosen because of their probability and severity as expected faults. A requirement concerning the tolerance with regard to a certain fault could be formulated as follows:

The system needs to be tolerant with the regard to the fault: continuous malfunction of sensor XYZ because of mechanical sensor destruction. In the case of this fault the main system performance indicators need to remain within the range between 80 and 100% of their value in the non-faulty case.

Following elements are part of this requirement:

- *The system needs to be tolerant with the regard to the fault*: general introduction;
- *continuous*: time behaviour of the expected fault;
- *malfunction of*: type of the expected fault;
- *sensor XYZ*: faulty component(s) and/or connection(s);
- *because of mechanical sensor destruction*: cause of the expected fault (optional);
- *In the case of this fault the main system performance indicators need to remain within the range between 80 and 100% of their value in the non-faulty case.*: expected minimum level of fault-tolerance.

Obviously, it is not sensible to restrict fault-tolerance completely to expected faults. Additionally, in early stages of design components may still be undefined and a decision for expected faults may be impossible. In this case, characteristic types of faults can be defined such as "type 1 fault: fault of one crucial component". In this case a requirement formulation could be:

The system needs to be tolerant with the regard to any imaginable single type 1 fault. In the case of this kind of fault the main system performance indicators need to remain within the range between 80 and 100% of their value in the non-faulty case.

In the scope of reliability, safety and fault-tolerance *redundancy* (i.e. the addition of some system elements, which function as backup in the case of fault or failure (compare [13]) is sometimes listed as the ultimate measure for improvement. However, redundancy can also lead to disadvantages, e.g. if an auxiliary unit causes additional cost, weight and space or if several smaller units are less efficient than one larger unit could be. Additionally, sometimes redundant units can be susceptible to the same kind of fault; in this case the improvement can be very small or even non-existing. The decision to apply redundancy for a technical system is very complex, because it is frequently a question of additional equipment and only sometimes it is necessary whereas sometimes redundancy is excessive [43]. The effect of redundancy can sometimes be smaller than expected, because two redundant systems may not be independent and in some events such as a counter overflow of a computer also a redundant computer may be subjected to the same fault (compare [13]). It is therefore necessary to take a closer look at the concept of redundancy also including concepts such as diversity (this will be done in the next sub-sections) and to consciously to describe the amount and type of redundancy in the requirements. It is also important to note that the quality of redundancy can be different on the different levels of abstraction (compare Fig. 3.4).

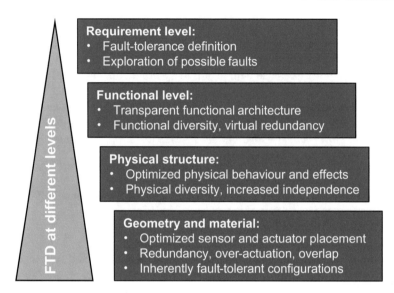

Fig. 3.4 FTD at different abstraction levels

The obvious forms of redundancy can be found on the most concrete level "geometry and material". One example could be two optical sensors observing the same scene. More independence can be achieved on the next higher level "physical structure". In the case of the example, now one optical sensor and another sensor relying on different physical effects, e.g. a ultrasonic sensor, are used. Even more independence could result from diversity on the functional level. One example would be to replace one real sensor by a virtual sensor, which works different even in the abstract functional domain. The possible measures to enhance fault tolerance on each level of abstraction will be explained in detail in the next sections.

The most prominent methods which can assist in the exploration of requirements are the "Fault Tree Analysis" (FTA), the "Failure Mode and Effects Analysis" (FMEA) and "benchmarking".

The "Fault Tree Analysis" (FTA) is a structured procedure for the identification of internal and external causes, which, whether they occur on their own or in combination, can lead to a faulty state of a technical system (compare [5]). This method is used worldwide in many different industries and allows quantifying the system reliability. The FTA is based on Boolean algebra and probability theory and can be applied as a diagnosis and development tool and is especially helpful in early design stages [5].

A "Failure Modes and Effects Analysis" (FMEA) is an engineering analysis performed by an interdisciplinary team of experts that thoroughly analyses product designs or manufacturing processes, early in the design process [8, 46]. The objective of an FMEA is finding and correcting weaknesses before the product gets into the hands of the customer but it can also be used in order to identify possible faults and the respective counter-action. Therefore a FMEA performed on successor systems or competitor systems can contribute to the exploration of requirements.

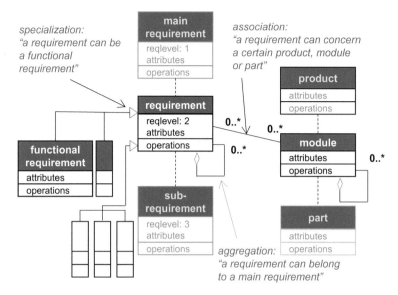

Fig. 3.5 General structure of model-based requirements management (compare [23])

Benchmarking in its core sense is a systematic process with mechanisms to measure, compare and evaluate products and processes. The main types of benchmarking are depending on the questions "what to benchmark" and "whom to benchmark against" (compare [6]). Current research mainly concerns benchmarking of the internal operations of companies; Albertin et al. name product benchmarking sometimes referred to as the first generation of benchmarking [2]. However, the in-depth analysis of competitive systems is still very common in industrial companies and can support the exploration of requirements. In the scope of fault-tolerance, a benchmarking of the reliability, safety and fault-tolerance of competitive technical systems can be extremely fruitful.

The exploration of requirements can result in an abundance of important requirements. In order to cope with this large number the strategies, methods and tools of requirements management were proposed. The ISO-Standard 29148 [28] defines requirements management as "activities that ensure requirements are identified, documented, maintained, communicated and traced through-out the life cycle of a system, product, or service". In industrial settings, the exploration of requirements is usually connected with some kind of requirements management; this is especially important in the case of reliability and safety requirements which need to be documented for certification and proof of good engineering practice purposes.

In recent years, an approach of model based requirements management based on graph-based languages has been developed [23]. Figure 3.5 shows the general structure of this approach.

In the innovative approach the modelling of requirements has in very general terms three main aspects which could also be understood as dimensions. Firstly, requirements can be divided into different domains (e.g. technical requirements). Secondly, a requirement can be part of a main-requirement or can itself lead to 2 or more sub-requirements. For instance, the main-requirement of a battery "store energy" can lead to the sub-requirements "provide current" and "endure loading cycles". Thirdly, requirements can be linked to modules or components of the technical system or to the technical system as a whole. The applicability of the approach was demonstrated using the design of a gear system for urban trains.

To summarise, the most important aspects of FTD on the level of requirements are:

- The exploration of requirements should include the possible faults, the expected faults, the required level of fault-tolerance and the required form and amount of redundancy.
- The exploration of requirements can be aided by FTA, FMEA and benchmarking as well as by model-based requirements management.

A substantial share of requirements concerns functions which need to be realised in the technical system; these will be discussed in the next section.

3.2 Functional Architecture

The first solution oriented level of the product concretization model (compare Fig. 3.3) is concerned with the functional architecture of a technical system. For conscious FTD it is necessary to have a transparent view of the functional architecture which can be provided by function models. Numerous research activities concern the development of function models; Eisenbart et al. [16] have complied an excellent review. Two main concepts of a function could be identified:

- A function describes the ability of a system to achieve a goal or fulfil a given task by showing certain behaviour.
- A function describes the transformation, conversion or change of states of distinct operands (i.e. typically specifications of material, energy or signals).

To conclude, a function model can be any kind of representation of one or more functions and logical relationships between them. In recent years, a function model including states, actors and hierarchies has been developed [40]. Figure 3.6 shows a function model of a wheel-hub motor for an electrical vehicle.

In the upper part of the figure the main flow—an energy flow—is visible with the operation "accelerate/decelerate wheel", the two states "wheel with actual speed" and "wheel with desired speed" and an operator "wheel-hub motor with speed sensing". One auxiliary flow—also an energy flow—is connected to the main flow by means of a condition state. This means that electrical energy is a condition for the operation in the main flow. This energy is regulated by a motor control unit. The second auxiliary flow—a signal flow—is connected to the main flow by means of a process state. This state—actual wheel speed—is generated during the operation in the main flow. On the boundary of the investigated system the actual vehicle velocity is provided e.g. by

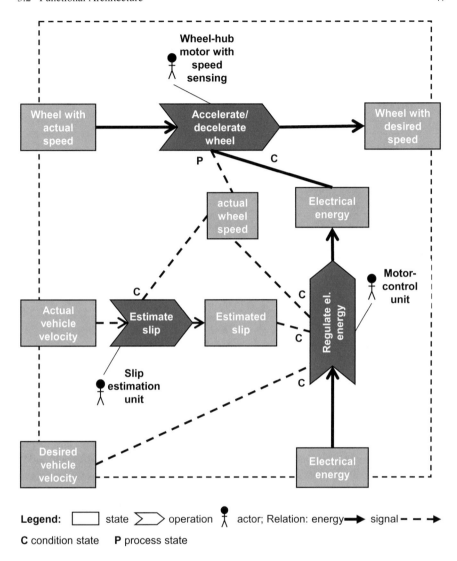

Fig. 3.6 Function model of a wheel hub motor

a GPS system. This information and the actual wheel speed can be used to estimate the wheel slip. Another information on the system boundary is the desired vehicle velocity which may come from the driver or a cruise control system. This information together with information about the actual wheel speed and the slip estimate allow the regulation of the wheel hub motor.

The function model allows an in-depth analysis of technical systems including mechanical, electrical and control components. This methodology was expanded by using graph-based design languages by Ramsaier et al. [41]. An example describing operations and actors in a multicopter is shown in Fig. 3.7.

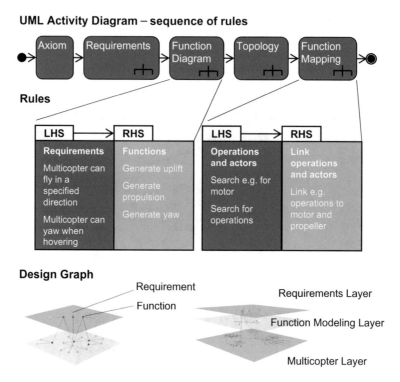

Fig. 3.7 Model-to-model transformations for functional modeling [41]

The upper part of Fig. 3.7 shows a sequence of so-called rules in a "Unified Modelling Language" (UML) activity diagram. In graph-based design languages, rules are formulated with model-to-model transformations. On the "Left Hand Side" (LHS) of a rule, the instances are formulated which the compilation engine should search within the design graph. The "Right Hand Side" (RHS) of the rule shows what transformations are to be done. The second rule after "requirements" is the subProgram "FunctionDiagram". This rule has on its left hand side the functional requirements and on its right hand side the functions consisting of operations and states. Then a rule is carried out which generates the topology of the multicopter. The subProgram at the end named "FunctionMapping" describes the mapping of functions to actors and has on its left hand side operations and actors independently and on its right hand side operations and actors sensibly linked. The lower part of Fig. 3.7 depicts concerned layers of the design graph. The first subProgram "FunctionDiagram" leads to the the two layers "Requirements Layer" and "Function Modelling Layer". The second subProgram "FunctionMapping" additionally leads to the "Multicopter Layer" which contains the aggregations of the components of the multicopter.

Besides the transparency of the functional architecture, which can be achieved by means of function models in connection with graph-based languages another important aspect concerns "functional redundancy". The term functional redundancy is frequently used in biology and denotes that more than one species share the same function in a ecosystem in a near neutral way [47]. In the field of robotics "functional redundancy" means that for the same trajectory, different robot configurations are possible [34]. Feng et al. [18] report about the development of a "multi-Sensor Error Detection" (SED) and "Functional Redundancy" (FR) system which relies on two different technologies, ("Outlier-Robust Kalman filter" (ORKF) and "Locally-Weighted Partial Least Squares" (LW-PLS) regression model). In the scope of FTD for technical systems "functional redundancy" can be defined as the application of two or more functionally different components (sensor, actors, process components) or algorithms in order to allow a higher level of fault-tolerance. In this scope "functionally different" means a considerably altered interplay of states and operations in a system. A prominent example with enormous potential is the use of analytical redundancy [25].

To conclude, the main aspects of FTD on the level of functional architecture are:

- An overview and understanding of the functional architecture of technical systems can be fostered by function models and model-based function modelling.
- Functional diversity is the most elaborate form of redundancy and can greatly enhance the fault-tolerance of technical systems, if consciously applied.

The physical realization of a technical system can be derived from function models.

3.3 Physical Realization

The intermediate level between the rather abstract functions and the concrete design parameters (geometry, etc.) is emphasised by several design methodologies. Sometimes it is referred to as "behaviour", i.e. what a design artefact does [19, 29, 53]. In the "Theory of Inventive Problem Solving" (TIPS), an analysis tools called "Su-Field analysis" is used to model a technical system. The basic idea of a Su-Field model is that any part of a technical system can be represented as a set of substance components and field interactions among these components [55]. In the German design methodology lists of physical effects are provided, which offer physical solutions for certain input and output parameters [15, 37, 38]. The application of these lists can widen the space of possible solutions. In general it can be stated that the engineers try to design an optimum physical behaviour for their technical systems. In order to increase the fault-tolerance, the probably most important concept for FTD on the physical level is physical diversity, i.e. the use of system elements which differ in the applied physical effects for achieving specific objectives. Current airplanes e.g. the Airbus A380 achieve their high level of fault-tolerance by means of a mixture of hydraulic power lines and hydraulic actuators together with electrical power lines and electrical actuators and are using physical diversity as economical means

Physical effect chain

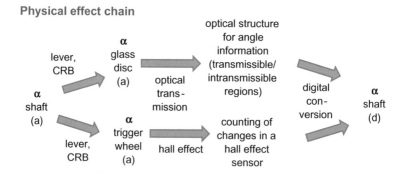

(a): analog; (d): digital; CRB: cohesion of rigid bodies

Fig. 3.8 Fault-tolerant physical effect chain

of fault-tolerance [9]. Physical diversity is also an important concept in the sensing systems of autonomous vehicles. Usually these vehicles use optical sensors (cameras) in order to capture their current environment. However, they cannot rely on optical sensors as only means e.g. in the case of fog. Therefore sensors which are using different physical effects such as "Light Detection And Ranging" (LIDAR) sensors are used in addition and the information is gathered by means of sensor fusion [10].

A detailed example describing a fault-tolerant physical effect chain is shown in Fig. 3.8 (compare [51]).

The current angle of the wheel α is present at the shaft. By means of the physical effects "lever" and "Cohesion of Rigid Bodies", machine elements transmit the angle to a glass disc and at the same time a trigger wheel. On the glass disc are certain structures of transmissible and intransmissible regions, which allow identifying the absolute angular position by means of the physical effect "optical transmission". The trigger wheel move close to a hall sensor allowing to count changes. Both pieces of information are then converted in digital angular information with regards to the angle of the wheel. In the case of a fault in one of the sensors it is still possible to detect the position of the shaft. Obviously, this kind of effect chain allows analysis and depiction of the physics acting in a product, thus fostering a deeper understanding and enabling communication between engineers [51].

To summarise the aspects of FTD on the level of physical realization:

- The conscious design of the physical behaviour of technical systems can be fostered by analysing the physical effects and lists with potential physical effects.
- Physical diversity is an important means to increase the independence of redundant system components and can thus enhance the fault-tolerance of technical systems, if consciously applied.

3.4 Geometrical Considerations

While, on the more abstract levels of FTD, a relatively small number of research publications can be found, a rich body of research is obvious on the level of concrete geometrical considerations. A large number of research works concern the geometrical placement of sensors and actuators.

The *placement of sensors* was intensively researched for decades in the scope of water distribution networks; an exhaustive review is presented by Rathi and Gupta [42]. Current work is concerned with adjoint-based numerical methods [54], the use of stoichiometry and kinetics of the reactions [36] and the application of hybrid feature selection [49]. Connected work concerns sensor placement for flood forecasting [17]. In the scope of the health monitoring of steel structures, Lu et al. [31] investigate optimal sensor placement and offer an review of related work. Current work also concerns sensor placement in thermo-mechanical systems such as machine tool columns [22].

The *placement of actuators* is in the scope of the research of Morris and Yang [33]; they define placement criteria for the control of structures. Irschik and Nader [26] investigate the placement of actuators for beams with piezoelectric actuation. In the scope of the research of Yunlong et al. [56] is the placement of actuators for vibration control systems for air planes.

A connection of sensor and actuator placement is researched in the field of power grids [52], flexible mechanical structures [20, 35] and leakage detection and localization [39].

It can be concluded that elaborate methods, algorithms and systems have been developed in order to support optimum sensor and actuator placement. Usually the research results focus on certain fields of application. They right choice of the geometrical placement of sensor and actuators remains a challenge for the designers of technical systems, especially if the fault-tolerance should be optimised.

Also on the most concrete level *redundancy* plays an important role—in this case the redundancy of system components which share the same functions and physical effects. Often on this level the fault-tolerance is enhanced by a duplication of elements. Two cases can be distinguished: hot and cold redundancy. In the case of hot redundancy, the duplicated elements are always together in operation such as the engines of an aircraft. In case of cold redundancy the additional system elements will only be in operation in case of a fault—such as auxiliary power generators of nuclear plants. Cold redundancy can have its limits, because elements can also become faulty in the time span when they are not operating. Such "latent" or "dormant" faults pose particular dangers because they can remain undetected [13].

Problems with *sensor redundancy* are also apparent. One major problem is caused by the fact that in some cases a sensor fault will not lead to *no* sensor reading but to a *wrong* sensor reading. In the case of two redundant sensors, the decision, which sensor reading to trust, is not a trivial one. One approach to tackle this problem are voting schemes, i.e. to trust the similar reading of at least two sensors. This does however require at least three redundant sensors. Another approach is using plausibility checks of sensor readings. The first level is to check whether the sensor reading is within a physically possible zone. The next level would be to check if the gradients of sensor readings are physically possible. Even more advanced would be the use of a virtual redundancy (compare Sect. 3.2). Today many sensors, especially sensors which deliver a digital signal, e.g. to a bus system, already dispose of such mechanisms. In this case, the sensors, which do not indicate a fault, can be trusted with a certain probability and the fault-tolerance can also be increased in the case of only two redundant sensors.

Also *actuator redundancy* can lead to more problems than originally expected. Blocked actuators may hinder the movement of certain parts, even if a second, redundant actuator is present and could theoretically perform the required action. In this case further measures are necessary, e.g. to provide mechanisms which are able to disconnect an actuator in the case of its fault.

Three other concepts are related to redundancy and can increase the fault-tolerance of technical systems: "over-actuation", "sensor-overlap" and "actuator-overlap".

Over-actuation is often understood as the usage of more actuators than necessary for controlling the rigid-body modes of motion systems [45, 48]. Another possibility for over-actuation is the use of stronger actuators than necessary [51]. Over-actuated designs usually have the advantage of a better controllability and can increase fault-tolerance, because the over-actuation potential can be used to compensate the effects of faults [51].

The notion *sensor-overlap* indicates that certain sensors which deliver not only a single value but 1D, 2D or 3D information such as optical sensors can be directed in a manner so that their sensing zones overlap. The overlapping zones can then be used in order to detect, if the sensors are giving at least similar information for these zones, and thus for concluding that they are working fault-free.

In some scenarios actuators can work collaboratively in order to achieve a given task, for example the legs in a six-leg walking robot. In such cases *actuator-overlap* is possible. In this case, one actuator can at least partly take over the tasks of another actuator and can thus increase fault-tolerance.

Elaborate methods exist in order to evaluate the reliability of technical systems with redundant elements; frequently they are based on fault tree analysis [3] or Monte Carlo Simulation [1].

Two strategies from "Design for Safety" (DfS) (compare [21]) should also be considered in the most concrete stage of fault-tolerant design. The strategy "Safe-Life" aims at systems which are by design rigid enough to sustain even extreme conditions. The complementary strategy is referred to as "Fail-Safe" and intends to avoid negative consequences in the case of the occurrence of a fault or failure. Some systems—such as the fuselage of an air plane—can only be designed "Safe-Life"; elaborate calculations and simulations as well as experiments are carried out in order to ensure that the respective system will survive unusual and even unexpected situations. This strategy will also lead to an increase of fault-tolerance, because certain situations that could theoretically lead to a fault which will, in this case, not lead to this fault. The second strategy "Fail-Safe" aims at enabling a controlled shut-down to a safe state when a critical fault occurs (compare [7]). The means to achieve this aim are similar to the means which allow accommodating faults, therefore the strategy "Fail-Safe" and the connected concepts, methods, algorithms and tools can also contribute to the fault-tolerance of technical systems.

An additional strategy of DfS is the design of inherently safe system configurations—a prominent example could be a compressor without a belt so no one could jam the fingers in this belt. This can be transferred to fault-tolerant design— a designer could intend to create "inherently fault-tolerant system configurations". One example could be an "Overhead Valve" (OHV) engine which, in contrast to an "Overhead Camshaft" (OHC) engine, does not dispose of a timing belt or chain. Thus, the faults "elongation of the timing chain" or "tearing of the timing belt" cannot occur (it is important to point at that besides this an OHV engine exhibits some disadvantages such as the high speed behaviour).

To summarise, the three most prominent aspects of FTD on the level of geometrical considerations can be listed:

- Elaborate methods, algorithms and systems exist, which allow an optimum geometrical placement of sensors and actuators also with regard to fault-tolerance.
- Generally, the fault-tolerance can be enhanced with redundant system elements. However, they cause less optimum systems in terms of economic value and performance and can still fail, especially in the case of cold redundancy. Therefore, the redundant elements need to be consciously chosen and concepts such as "over-actuation" and "overlap" need to be taken into consideration.
- The strategies of "Design for Safety" (DfS) such as "Safe-Life" and "Fail-Safe" can also increase fault-tolerance; an adopted strategy "inherently fault-tolerant system configurations" can also be applied for this purpose.

3.5 Summary

The aspects and components of fault-tolerant design were in the focus of this chapter. The discussion followed a widely accepted model of the levels of abstraction of a technical system. Together with the aspects and components of fault-tolerant control they can support design and control engineers in their integrated task to synthesise systems which dispose of increased fault-tolerance with only small additional expenditures, and, if possible, with uncompromised system performance.

References

1. Alban, A., Darji, H., Imamura, A., Nakayama, M.K.: Efficient Monte Carlo methods for estimating failure probabilities. Reliab. Eng. Syst. Saf. **165**, 376–394 (2017)
2. Albertin, M.R., Pontes, H.L.J., Frota, E.R., Assuncao, M.B.: Flexible benchmarking: a new reference model. Benchmarking Int. J. **22**(5), 920–944 (2015)

3. Bennett, J.W., Mecrow, B.C., Atkinson, D.J., Atkinson, G.J.: Safety-critical design of electromechanical actuation systems in commercial aircraft. IET Electr. Power Appl. **5**(1), 37–47 (2011)
4. Bernard, R., Irlinger, R.: About watches and cars: winning R and D strategies in two branches. In: International Symposium "Engineering Design The Art of Building Networks" (2016)
5. Bertsche, B.: Reliability in Automotive and Mechanical Engineering. Springer, Berlin (2008)
6. Bhutta, K.S., Huq, F.: Benchmarking best practices: an integrated approach. Benchmarking Int. J. **6**(3), 254–268 (1999)
7. Blanke, M., Frei, C.W., Kraus, F., Patton, R.J., Staroswiecki, M.: What is fault tolerant control? In: Proceedings of IFAC Symposium on Fault Detection Supervision and Safety of Technical Processes, SAFEPROCESS (2000)
8. Carlson, C.S.: Effective FMEAs: Achieving Safe, Reliable, and Economical Products and Processes using Failure Mode and Effects Analysis. Wiley, New York (2012)
9. Charrier, J.-J., Kulshreshtha, A.: A electric actuation for flight and engine control system: evolution, current trends and future challenges. In: Proceedings of the 45th AIAA Aerospace Sciences Meeting and Exhibit (2007)
10. Chavez-Garcia, R.O., Aycard, O.: Multiple sensor fusion and classification for moving object detection and tracking. IEEE Trans. Intell. Transp. Syst. **99**, 1–10 (2015)
11. Crnkovic, I., Asklund, U., Persson-Dahlqvist, A.: Implementing and Integrating Product Data Management and Software Configuration Management. Artech House, London (2003)
12. Daenzer, W.F., Huber, F.: Systems Engineering Methodik und Praxis. Verlag industrielle Organisation, Zurich (2002)
13. Downer, J.: When failure is an option: redundancy, reliability and regulation in complex technical systems. Centre for Analysis of Risk and Regulation (2009)
14. Dylla, N.: Denk- und Handlungsablufe beim Konstruieren. Hanser, Wien (1991)
15. Ehrlenspiel, K., Meerkamm, H.: Integrierte Produktentwicklung. Zusammenarbeit. Carl Hanser Verlag, Munich, Denkabläufe, Methodeneinsatz (2013)
16. Eisenbart, B., Gericke, K., Blessing, L.T.M., McAloone, T.C.: A dsm-based framework for integrated function modelling: concept, application and evaluation. Res. Eng. Des. **28**(1), 25–41 (2016)
17. Optimal sensors placement for flood forecasting modelling: Fattorusoa, G., Agrestab, A., Guarnieria, G., Lanzaa, B., Buonannoa, A., Molinarac, M., Marroccoc, C., De Vitoa, S., Tortorellac, F., Di Franciaa, G. Procedia Eng. **119**, 927–936 (2015)
18. Feng, J., Hajizadeh, I., Cinar, A., Samadi, S., Sevil, M., Frantz, N., Lazaro, C., Maloney, Z., Yu, X., Littlejohn, E., Quinn, L.: A multi-sensor error detection and functional redundancy algorithm for dynamic systems. In: Proceedings of the 2017 AIChE Annual Meeting (2017)
19. Gero, J.S., Kannengiesser, U.: The function-behaviour-structure ontology of design. In: Chakrabarti, A., Blessing, L.T.M. (eds.) An Anthology of Theories and Models of Design, pp. 263–283. Springer, Berlin (2014)
20. Gney, M., Eskinat, E.: Optimal actuator and sensor placement in flexible structures using closed-loop criteria. J. Sound Vib. **312**, 210–233 (2008)
21. Gullo, L.J., Dixon, J.: Design for Safety. Wiley, New York (2017)
22. Herzog, R., Riedel, I., Ucinski, D.: Optimal sensor placement for joint parameter and state estimation problems in large-scale dynamical systems with applications to thermo-mechanics. Technische Universitaet Chemnitz (2017)
23. Holder, K., Zech, A., Ramsaier, M., Stetter, R., Niedermeier, H.-P., Rudolph, S., Till, M.: Model-based requirements management in gear systems design based on graph-based design languages. Appl. Sci. **7**, (2017)
24. Hruschka, P.: Business Analysis und Requirements Engineering: Produkte und Prozesse nachhaltig verbessern. Hanser, Munich (2014)
25. Hu, B., Seiler, P.: A probabilistic method for certification of analytically redundant systems. Int. J. Appl. Math. Comput. Sci. **25**(1), 103–116 (2015)

26. Irschik, H., Nader, M.: Actuator placement in static bending of smart beams utilizing Mohr's analogy. Eng. Struct. **31**, 1698–1706 (2009)
27. Isermann, R.: Fault Diagnosis Systems. An Introduction from Fault Detection to Fault Tolerance. Springer, New York (2006)
28. ISO/IEC/IEEE 29148:2011: Systems and software engineering - Life cycle processes - Requirements engineering
29. Li, L., Yu, S., Tao, J., Li, L.: A FBS-based energy modelling method for energy efficiency-oriented design. J. Clean. Prod. **172**, 1–13 (2018)
30. Lindemann, U.: Methodische Entwicklung technischer Produkte. Springer, Berlin (2009)
31. Lu, W., Wen, R., Teng, J., Li, X., Li, C.: Data correlation analysis for optimal sensor placement using a bond energy algorithm. Measurement **91**, 509–518 (2016)
32. Morkos, B., Mathieson, J., Summers, J.D.: Comparative analysis of requirements change prediction models: manual, linguistic, and neural network. Res. Eng. Des. **25**, (2014)
33. Morris, K., Yang, S.: Comparison of actuator placement criteria for control of structures. J. Sound Vib. **353**, 1–18 (2015)
34. Mousavi, S., Gagnol, V., Bouzgarrou, B.C., Ray, P.: Stability optimization in robotic milling through the control of functional redundancies. Robot. Comput. Integr. Manuf. **50**, 181–192 (2018)
35. Nestorovic, T., Trajkov, M.: Optimal actuator and sensor placement based on balanced reduced models. Mech. Syst. Signal Process. **36**, 271–289 (2013)
36. Ohar, Z., Lahav, O., Ostfeld, A.: Optimal sensor placement for detecting organophosphate intrusions into water distribution systems. Water Res. **73**, 193–203 (2015)
37. Pahl, G., Beitz, W., Feldhusen, J., Grote, K.H.: Engineering Design: A Systematic Approach. Springer, Berlin (2007)
38. Ponn, J., Lindemann, U.: Konzeptentwicklung und Gestaltung technischer Produkte. Springer, Berlin (2011)
39. Przystalka, P., Moczulski, W.: Optimal placement of sensors and actuators for leakage detection and localization. In: Proceedings of the 8th IFAC Symposium on Fault Detection, Supervision and Safety of Technical Processes (SAFEPROCESS) (2012)
40. Ramsaier, M., Spindler, C., Stetter, R., Rudolph, S., Till, M.: Digital representation in multicopter design along the product life-cycle. Procedia CIRP **62**, 559–564 (2016)
41. Ramsaier, M., Stetter, R., Till, M., Rudolph, S., Schumacher, A.: Automatic definition of density-driven topology optimization with graph-based design languages. In: Proceedings of the 12th World Congress on Structural and Multidisciplinary Optimisation (2017)
42. Rathi, S., Gupta, R.: Sensor placement methods for contamination detection in water distribution networks: a review. Procedia Eng. **89**, 181–188 (2014)
43. Rogova, E.S.: Reliability assessment of redundant safety systems with degradation. Delft University of Technology, 2017
44. Rouissi, F., Hoblos, G.: Fault tolerant sensor network design with respect to diagnosability properties. In: Proceedings of the 8th IFAC Symposium on Fault Detection, Supervision and Safety of Technical Processes (SAFEPROCESS), pp. 1120–1124 (2012)
45. Ryll, M., Buelthoff, H.H., Giordano, P.R.: Overactuation in UAVs for enhanced aerial manipulation: a novel quadrotor concept with tilting propellers. In: Proceedings of the 6th International Workshop on Human-Friendly Robotics (2013)
46. SAE J 1739:2009: Potential failure mode and effects analysis in design (design FMEA) and potential failure mode and effects analysis in manufacturing and assembly processes (Process FMEA) and effects analysis for machinery (Machinery FMEA)
47. Scheffer, M., Vergnon, R., van Nes, E.H., Cuppen, J.G.M., Peeters, E.T.H.M., Leijs, R., Nilsson, A.N.: The evolution of functionally redundant species; evidence from beetles. PLOS ONE **10**(10), (2015)
48. Schneider, M.G.E., van de Molengraft, M.J.G., Steinbuch, M.: Benefits of over-actuation in motion systems. In: Proceeding of the 2004 American Control Conference (2004)
49. Soldevila, A., Blesa, J., Tornil-Sin, S., Fernandez-Canti, R.M., Puig, V.: Sensor placement for classifier-based leak localization in water distribution networks using hybrid feature selection. Comput. Chem. Eng. **108**, 152–162 (2018)

50. Stetter, R., Pulm, U.: Problems and chances in industrial mechatronic product development. In: Proceedings of the 17th International Conference on Engineering Design (ICED 09), vol. 5, pp. 97–108 (2009)
51. Stetter, R., Simundsson, A.: Design for control. In: Proceedings of the 21st International Conference on Engineering Design (ICED 17), vol. 4, Design Methods and Tools, pp. 149–158 (2017)
52. Summers, H.H., Lygeros, J.: Optimal sensor and actuator placement in complex dynamical networks. In: Proceedings of the 19th World Congress The International Federation of Automatic Control (2014)
53. Umeda, Y., Ishii, M., Yoshioka, M., Shimomura, Y., Tomiyama, T.: Supporting conceptual design based on the function-behavior-state modeler. Artif. Intell. Eng. Des. Anal. Manuf. AIEDAM **10**(4), 275–288 (1996)
54. Waeytens, J., Mahfoudhi, I., Chabchoub, M.-A., Chatellier, P.: Adjoint-based numerical method using standard engineering software for the optimal placement of chlorine sensors in drinking water networks. Environ. Model. Softw. **92**, 229–238 (2017)
55. Yan, F., Dridi, M., El Moundi, A.: An autonomous vehicle sequencing problem at intersections: a genetic algorithm approach. Int. J. Appl. Math. Comput. Sci. **23**(1), 183–200 (2013)
56. Yunlong, L., Xiaojun, W., Ren, H., Zhiping, Q.: Actuator placement robust optimization for vibration control system with interval parameters. Aerosp. Sci. Technol. **45**, 88–98 (2015)

Part II
Fault-Tolerant Design and Control
of Automated Vehicles

Chapter 4
Methodical and Model-Based Design of Automated Vehicles

This chapter describes all phases of a conscious design process of an automated guided vehicle with a special consideration of the fault-tolerance of this system. Special emphasis is placed both on the methodical aspects of the conscious design process and on digital models which support and organise all stages of the design process. The chapter starts with Sect. 4.1 which describes a distinct level of a design process—the planning and control of this process which are ongoing endeavours. The early phases of a design process concern the exploration of customer needs and are described in Sect. 4.2. The requirements management, which is the content of Sect. 4.3, is ongoing throughout the process but has a point of main emphasis in the earlier stages. Section 4.4 reports methods and models of the interdisciplinary stage *system design*, whereas the *domain specific design* is discussed in Sect. 4.5. The stage *system integration* is explained in Sect. 4.6 and the different levels of *verification and validation* are discussed in Sect. 4.7. The contents of this chapter are based on the publications [28, 52].

4.1 Process Planning

The product development of complex systems such as Automated Guided Vehicles (AGVs) requires certain mechanisms in order to achieve interconnected, manageable and organised processes. One prominent means to support engineers and design managers in this challenge are process flow models. Such models depict certain phases, stages or steps which allow to decompose the complete design process and to describe a certain design procedure. Frequently, these models try to represent the underlying logic of a product development process. In recent years, model-driven approaches towards process planning and control have found rising attention [46]. It is important to note that the process models are not only used to arrange the phases, stages or steps in a logical order, but also to allow time for planning and the assignment of responsibilities and activities to individual persons or certain departments of the

© Springer Nature Switzerland AG 2020
R. Stetter, *Fault-Tolerant Design and Control of Automated Vehicles and Processes*, Studies in Systems, Decision and Control 201,
https://doi.org/10.1007/978-3-030-12846-3_4

product development organization. It can be concluded that process models can assist several aspects of the conscious management of design. The central aspects concerning fault-tolerant design and control are the planning of phases for a conscious search for possible and probable faults, phases for the development of fault-tolerant control algorithms and phases for the verification of the mechanisms for increased fault-tolerance.

In the scientific communities of systematic design, new product development, software development and systems engineering, a multitude of process models has been developed in the last decades. Some prominent examples are explained in the VDI guidelines 2221: "Methodology for development and design of technical systems and products" [57] and 2422: "Design procedure for mechanical devices with microelectronics control" [58]. Since more than one decade a consensus can be observed that a process model called V-model (compare Chap. 3) is suitable for representing the general logic of the product development process of mechatronic systems. The VDI guideline 2206 "Design methodology for mechatronical systems" describes this process model [25, 56]. A large body of research has covered the application, optimization and expansion of the methodology [1, 2, 5, 6, 10, 13, 15, 31]; it has been developed into a standard process model in industrial companies that develop complex mechatronic products such as AGVs.

Complex mechatronic products consist of mechanical, electrical and electronic subsystems and apply computer software for their functioning and additional purposes (e.g. service, surveillance, etc.). The V-model explicitly addresses this interdisciplinary engineering discipline. The underlying logic of this model aims at integrating the processes, models and knowledge of mechanical engineering, electrical engineering and computer science.

One main focus of the V-model is the interdisciplinary system design. System engineers and engineers of the different disciplines need to cooperate in order to assign all required functions to solutions which can realise these functions in an optimum manner with the smallest expenses, irrespective of the concerned disciplines. The V-model also acknowledges that certain detail activities still have to be carried out by the specialist of the three disciplines, but clarifies that this separation necessitates a conscious integration of the domain-specific solutions. Furthermore, the V-model points out the necessity of verification and validation activities such as simulation and testing on different levels of detail and different product integration levels. The structure of the V-model is shown in Fig. 4.1.

In contrast to other publications, a distinct phase *customer needs* is added at the beginning of the process; the significance of this phase is addressed in Sect. 4.2. The final product of the process is a complete description of the product consisting of data and prototypes allowing the production of the product.

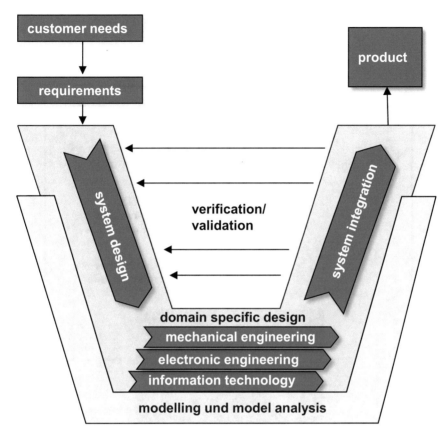

Fig. 4.1 General structure of the V-model

4.1.1 Development Methodology for Mechatronic Systems

The VDI guideline 2206 does not only describe the V-model as a process model, but also proposes the general problem-solving cycle, but on a different solution level. The V-model is proposed as procedure scheme on the macro level, i.e. for planning and controlling complete product development processes or at least larger components of such processes. The problem solving cycle is proposed as general procedure scheme on the micro level, i.e. for planning and controlling smaller segments of the product development processes which are dedicated for solving a certain problem. Such segments can last from few hours up to several months. The origin of the problem solving cycle can be found in systems engineering [21]. The general structure of the problem solving cycle is shown in Fig. 4.2.

This procedure scheme consists of several steps which are arranged in a logical order. The start of a problem solving cycle are the steps *situation analysis* or *adoption of a goal*, depending of the problem situation. If an existing structure is available and

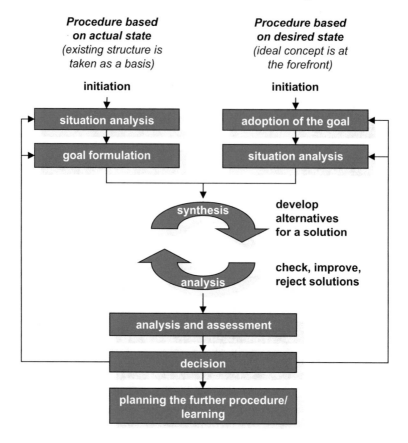

Fig. 4.2 The problem-solving cycle

can be analysed, then it is sensible to start with an in-depth analysis of the situation. A concentration on the goals can be advantageous, if the generation of an ideal solution from the scratch is desirable. The core of the scheme is a continuous cycle between synthesis (creating new solutions) and analysis (evaluating characteristics of these solutions). The *analysis and assessment* step evaluates explicitly how good certain solutions can fulfil the given goals. The last step *decision* is based on the results of this evaluation, but also takes feelings of human beings into consideration.

As stated above, the V-model can serve as a procedure scheme for complete product development processes. For complex systems it can also be sensible to repeat this procedure several times on different levels of maturity (compare Fig. 4.3).

The detailed planning of a product development process is depending on many aspects of the product itself, but also the product development environment. In the last decade, approaches for a situation specific planning of product development process have been proposed, for instance the "Product Model Driven Development" PMDD [46]. For mechatronic product development processes a function oriented planning and synchronization have been recommended [27].

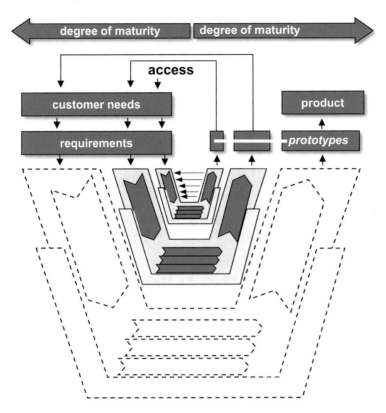

Fig. 4.3 Proceeding through a number of macro-cycles

In recent years, agile methods have found growing attention. Especially in software engineering, the framework "Scrum" is frequently applied in order to enable agile development processes (the term "Scrum" is an expression from the sport rugby for a very close conglomeration of the teams). Working in this framework is characterised by a team which is performing development work in so-called sprints which usually have the duration of one week. During the sprint the team only has a short daily meeting (daily scrum) and during the remaining time the team members can concentrate of the development work [48]. They are assisted by a scrum master who tries to isolate the team from distracting factors. The requirements are brought into the process by the product owner. Before a sprint, a sprint planning meeting allows to clarify the scope and goals of a sprint. After the sprint, the results are presented in a sprint review. Figure 4.4 sketches the basic structure of the framework Scrum.

In contrast to traditional development methods, the development team is granted more self-organization. This includes, among other things, deciding on which tasks the developers will work next, or deciding within the own team how they will organise the implementation of the requirements [38]. Central advantages of Scrum are:

- the direct communication within the team,
- the agile/flexible proceeding because of adaptive planning before each sprint,

Fig. 4.4 The framework "Scrum"

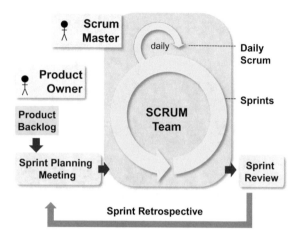

- an efficient process because of self-organization,
- during a sprint the team members can concentrate on the development tasks and are not distracted, for instance, by management interferences,
- in the sprint review also the process can be reflected thus leading to continuous improvement cycles.

In spite of its many advantages, Scrum cannot directly be applied in product development processes for mechatronic products which also involve physical components and physical testing, because the realization times and the complexity of the planning processes prohibit a completely agile development process (compare [43]). Currently, design science is looking at possibilities to apply a refined and expanded version of this framework also for the development of mechatronic products [34, 35].

In addition to the presented procedure schemes on the micro- and macro level, a number of methods and tools can be used to support the planning and control of the design process; such methods and tools and described in the next section.

4.1.2 Planning and Control of the Development Process

The planning and controlling of the product development process of complex mechatronic products such as AGVs is a major challenge, because the interconnections and interdependencies between the different subsystems and components are manifold and the determination of a sensible development sequence is anything but trivial. Serious problems can arise, if central components have to be changed late in the process, because these changes can lead to numerous necessary changes in other components. It is therefore sensible especially in the design of complex mechatronic systems to identify the central components, i.e. the components which can influence many other components. Such influences frequently are interdisciplinary.

Fig. 4.5 Influence matrix of
an AGV (simplified
example)

		influence on				
Which subsystems have an influence on which subsystems?		chassis	steering sys.	control system	drive system	active sum
influence of	chassis		0	1	1	2
	steering sys.	1		2	0	3
	control system	2	2		1	5
	drive system	2	1	3		6
	passive sum	5	3	6	2	

One example might be that certain kinds of electrical motors need to be controlled by certain motor control systems. One prominent method which can be used for analysing the degree of influence between subsystems or components is the influence matrix. This matrix consists of rows and columns with the subsystems or components. In the center of the matrix is an assessment of the degree of influence of one subsystem or component to another subsystem or component. These assessments are summarised in rows as passive sum and the columns add up as active sum. A simplified influence matrix of an AGV is shown in Fig. 4.5.

One advantageous depiction of the outcome of an influence matrix is a portfolio. This depiction differentiates four different kinds of subsystems or components:

- *Buffering* subsystems or components neither influence many other subsystems or components nor are influenced by many other subsystems or components.
- *Active* subsystems or components can influence a number of subsystems or components which is larger than the number of subsystems or components which can influence them.
- *Passive* subsystems or components are influenced by a number of subsystems or components which is larger than the number of subsystems or components which they can influence.
- *Critical* subsystems or components can both influence a large number of subsystems or components and can be influenced by a large number of subsystems or components.

In Fig. 4.6 a simplified influence portfolio of an AGV is shown, which is the outcome of the exemplary influence matrix.

For effective and efficient product development processes, critical and active subsystems or components need to be defined early in the process. The buffering and active subsystems or components can be defined later, because, if they are changed, no subsequent necessary changes of other systems or components are to be expected. This assessment cannot be limited to single disciplines, but has to include mechanical, electric and electronic subsystems and components and software. It can be one cornerstone to successful design of complex systems to identify the influential subsystems and components and to derive a sensible product development sequence.

Fig. 4.6 Influence portfolio
of an AGV (simplified
example)

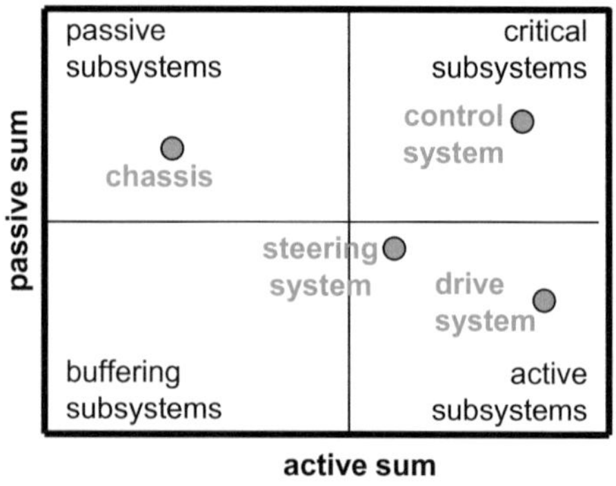

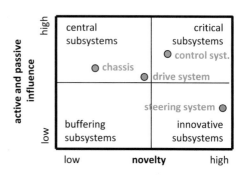

Fig. 4.7 Project sequence
portfolio of an AGV
(simplified example)

A detailed analysis of several product development processes led to the insight that
the degree of novelty of a subsystem or a component is another prominent characteristic which is important for the sequence of a design process. Usually a novel system
is composed of known, adapted and new components and subsystems. Obviously,
the highest risk is present in the new subsystems and components. Consequently, for
stable and manageable design processes, it is sensible to develop these subsystems
and components first, because each of them can cause a failure of the project. Possible
criteria for identifying new subsystems and components are the degree of innovation, the maturity of the development methodology, the simulation possibilities and
the achieved degree of fault-tolerance. It is possible to combine the characteristic
influence and *novelty* in one project sequence portfolio (compare Fig. 4.7).

Based on this portfolio, a design sequence can be developed which allows a
concentration of the resources to the ultimately critical subsystems and components.
One criterion for these critical subsystems and components is a low level of fault-
tolerance. The developed design sequence can be one important basis for an effective
planning and control of the design process.

For complex mechatronic products, a compromise between agility and structure
can be achieved by using an achievement oriented stage-gate process as shown in
Fig. 4.8.

The process consists of phases and stage-gates and is supported by maintaining
a continuously expanding requirements model. Similar to Scrum, each phase starts
with a fine planning. This fine planning concentrates on the achievements which this

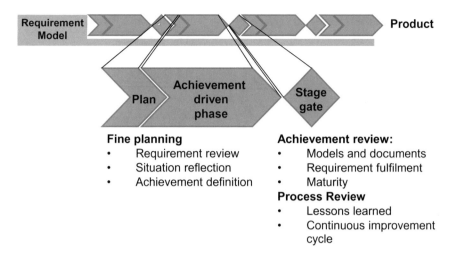

Fig. 4.8 Achievement oriented stage-gate process

phase should result in—no attempt is made to define in detail the activities in the phase. This achievement definition is based on a requirement review and a situation reflection. Typical achievements would be, for instance, product models with a certain degree of detail, completeness and maturity or testing results or realised production equipment such as product-specific tools. Each phase ends with a scheduled stage-gate. This gate is characterised by a review consisting of an achievement review and a process review. The achievement review concerns:

- models and documents and their degree of detail, completeness and maturity,
- a prognosis of the requirement fulfilment of the technical system under development based on estimations of experts and triangulation,
- an estimation of the maturity of the technical system under development based on estimations of experts and triangulation.

The process review consists mainly of a lessons learned session with the intention of leading to a continuous improvement cycle.

In general, the implementation of "Fault-Tolerant Approaches" (FTAP), i.e. guidelines, algorithms, tools, methods and strategies which aim at increasing the fault-tolerance of technical systems, in industry is a challenging task, which can be compared to the implementation of methods [51]. For the implementation of methods, a distinction of logical layers was developed [49], which can be adapted to the implementation of FTAPs. The distinct layers are:

- Initiation of the FTAP implementation process: this layer contains the activities that have to be performed in the very early stages of a FTAP implementation process.
- Analysis of the product development system: this layer contains activities aimed at creating an in-depth understanding of the technical system and the development processes.

Implementation Layers   **Success Factors**

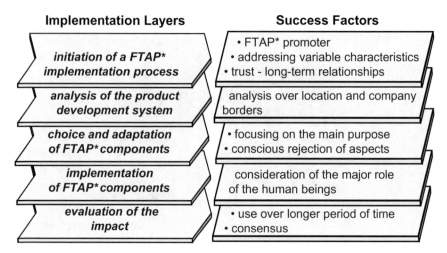

*) **FTAP**: Fault-Tolerant Approach
            including algorithms, guidelines, methods, etc.

Fig. 4.9 Implementation layers and success factors for fault-tolerant approaches

- Choice and adaptation of FTAP components: this layer describes the conscious choice of appropriate FTA components and the adaptation to the given product and process environment.
- Implementation of FTAP components: the core of this layer are means to mediate the objectives and the underlying logic of FTAP components as well as the skills to apply them effectively.
- Evaluation of the impact: this layer contains activities intended to evaluate the impact of a FTAP implementation, i.e. to determine the effect of the FTAP components. This evaluation is, amongst others, necessary to realise a continuous improvement cycle.

The implementation layers for FTAPs are shown in Fig. 4.9 (left side).

A collection of success factors for method implementation was gathered over the last decades; these insights can be adapted to the implementation of FTAPs. The chance that a fault-tolerant approach will be used in the longer term can be increased, if the approach was initially promoted by someone deeply convinced of the benefits. An implementation may be successful when the variable characteristics were addressed, i.e. the characteristics of the product development system which actually could be changed, because attempts to change characteristics outside the scope of change of the implementation team are frequently doomed to failure. A central success factor can be trust—the engineers need to trust external sources; this can be fostered by long-term relationships between industry and academia. Today, the processes can only be fully understood, if the process segments are analysed which take place at the engineering consultant and supplier companies. For successful implementation of FTAPs, overambitious attempts should be avoided and less important issues should be consciously rejected. One should not neglect that the knowledge and attitude of the involved engineers play a major role. Also in the evaluation of the impact, the use should be observed over a longer period of time and a consensus concerning the impact should be achieved. These success factors are summarised in Fig. 4.9 (right side).

4.2 Customer Needs Exploration

In many industrial branches, customer orientation is becoming increasingly important, because of a number of industrial trends such as the fact that many markets are saturated, i.e. products and services are increasingly interchangeable. Another general trend is an increased customer interaction, because customers are becoming increasingly critical. For successful companies the price of a product cannot be the only selling point. It is therefore very important for companies to explore the customer needs. Essential concepts of this exploration are described in the so-called Kano model (named after its Japanese inventor Noriaki Kano) [33]. This model helps to integrate the customer needs in the subsequent phases of a product development process [55]. The Kano model describes the relationship between customer needs and customer satisfaction. Customer expectations concerning the features and characteristics of products, systems, solutions or software vary widely. One feature could inspire one customer, the next customer might take the same feature for granted, and other customers might reject the product just because of this feature. Figure 4.10 shows the well-known kano model.

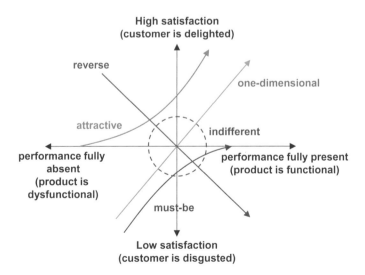

Fig. 4.10 Kano model

In this model five types of product attributes are visible:

- The absence of a *must-be* attribute will lead to costumer dissatisfaction, but it cannot lead to more customer satisfaction. For an AGV the functional safety is today a *must-be* attribute.
- *One-dimensional* attributes combine their fulfilment with customer satisfaction. The velocity of an AGV can be a *one-dimensional* attribute.
- *Attractive* attributes produces higher customer satisfaction, but they are not required in the product. For an AGV a modern design might be an *attractive* attribute.
- The costumer satisfaction is not affected by *indifferent* attributes; one example could be the colour of an AGV.
- *Reverse* attributes combine their fulfilment with low customer satisfaction; one example could be the corrosion of parts of an AGV.

In general these attributes can be used in order to structure the voice of the customer. AGVs are usually bought be producing or infrastructure companies. Their main interest is to make profit; therefore the most prominent customer needs concern the financial side of the application of AGVs. This can, however, not be reduced to the selling price of an AGV—the main aspects are usually its performance and efficiency. It is necessary incorporate many aspects into the comparison of AGV solutions. One possibility is the formulation of a Comparative Customer Value CCV.

Equations (4.1)–(4.3) allow a conscious consideration of the financial customer needs concerning an AGV solution. Equation (4.1) can be used in order to determine the customer comparative value CCV.

$$CCV = W_P * P + W_A * A + W_{IC} * IC + W_{OC} * OC \qquad (4.1)$$

where

- CCV denotes the customer comparative value of a certain solution which can be used to compare different AGV solutions;
- W_P denotes the weight of the performance for the specific logistic scenario on a scale from 0 to 1;
- P denotes an assessment of the expected performance on a scale from 0 to 1; 0 meaning no performance at all, 0.9 characterizing a performance which fully satisfies the respective logistic scenario and 1.0 characterizing a performance which is better than satisfactory but will still give some further advantages. P can also be determined using Eq. (4.2);
- W_A denotes the weight of the availability for the specific logistic scenario on a scale from 0 to 1;
- A denotes an assessment of the expected availability on a scale from 0 to 1; 0 indicating a very low availability (e.g. a very unreliable system) and 1.0 characterizing a system which is always available;
- W_{IC} denotes the weight of the investment costs for the specific logistic scenario on a scale from 0 to 1;

- IC denotes an assessment of the expected investment costs on a scale from 0 to 1; 0 indicating a very expensive solution (in terms of investment costs) and 1.0 a very economical (in terms of investment costs);
- W_{OC} denotes the weight of the operating costs for the specific logistic scenario on a scale from 0 to 1;
- OC denotes an assessment of the expected operating costs on a scale from 0 to 1; 0 indicating a solution which is very expensive in terms of operating costs and 1.0 a solution which is very economical in terms of operating costs. OC can also be determined using Eq. (4.3);
- the weights W_P, W_A, W_{IC}, W_{OC} have to add up to 1.0.

The assessment of the performance can be enhanced using the following equation:

$$P = W_{vel} * vel + W_{acc} * acc + W_M * M + W_{CC} * CC + W_{RR} * RR \quad (4.2)$$

where

- P denotes the performance of a certain solution which is used in Eq. (4.1);
- W_{vel} denotes the weight of the velocity for the specific logistic scenario on a scale from 0 to 1;
- vel denotes an assessment of the expected velocity on a scale from 0 to 1; 0 meaning very slow, 0.9 characterizing a velocity which fully satisfies the respective logistic scenario and 1.0 characterizing a velocity which is better than satisfactory but will still give some further advantages;
- W_{acc} denotes the weight of the acceleration for the specific logistic scenario on a scale from 0 to 1;
- acc denotes an assessment of the expected acceleration on a scale from 0 to 1; 0 indicating a very low acceleration and 1.0 characterizing a system with very high acceleration capabilities;
- W_M denotes the weight of the manoeuvrability for the specific logistic scenario on a scale from 0 to 1;
- M denotes an assessment of the expected manoeuvrability on a scale from 0 to 1; 0 indicating a very limited manoeuvrability and 1.0 a superior manoeuvrability;
- W_{CC} denotes the weight of the carrying capacity for the specific logistic scenario on a scale from 0 to 1;
- CC denotes an assessment of the expected carrying capacity on a scale from 0 to 1; 0 indicating a solution with very low carrying capacity and 1.0 a solution with very high carrying capability;
- W_{RR} denotes the weight of the room requirements for the specific logistic scenario on a scale from 0 to 1;
- RR denotes an assessment of the expected room requirements on a scale from 0 to 1; 0 indicating a solution with very high room requirements and 1.0 a solution with very small room requirements;
- the weights W_{vel}, W_{acc}, W_M, W_{CC}, W_{RR} have to add up to 1.0.

Furthermore, the assessment of the operating costs can be enhanced using the following equation:

$$OC = W_{EC} * EC + W_{SC} * SC + W_{SurC} * SurC \tag{4.3}$$

where

- OC denotes the operating costs of a certain solution which are used in Eq. (4.1);
- W_{EC} denotes the weight of the energy costs for the specific logistic scenario on a scale from 0 to 1;
- EC denotes an assessment of the expected energy costs on a scale from 0 to 1; 0 indicating a very expensive solution (in terms of energy costs for the operation) and 1.0 a very economical solution (in terms of energy costs);
- W_{SC} denotes the weight of the service costs for the specific logistic scenario on a scale from 0 to 1;
- SC denotes an assessment of the expected service costs on a scale from 0 to 1; 0 indicating a very expensive solution (in terms of service costs, e.g. requiring frequent and expensive maintenance operations) and 1.0 a very economical solution (in terms of service costs);
- W_{SurC} denotes the weight of the surveillance costs for the specific logistic scenario on a scale from 0 to 1;
- $SurC$ denotes an assessment of the expected surveillance costs on a scale from 0 to 1; 0 indicating a very expensive solution (in terms of surveillance costs, e.g. a solution that requires continuous surveillance from a highly qualified human operator) and 1.0 a very economical solution (in terms of surveillance costs, e.g. a completely independent system which can assure complete availability and safety);
- the weights W_{EC}, W_{SC}, W_{SurC} have to add up to 1.0.

The Eqs. (4.1)–(4.3) can lead to technical requirements describing the goals which the AGV solution should achieve.

In recent years the customer needs exploration has been focusing on user experience. Information about users can be gathered through traditional market research methods (e.g. interviews at home, telephone surveys, etc.) and, especially for innovative products, by means of "product clinics"/"user clinics" [16]. New technologies such as Virtual Reality (VR) and Augmented Reality (AR) even allow to gather information about user experience from future products which only exist virtually. Figure 4.11 summarises the different aspects concerning user experience.

4.3 Requirements Management

The special role of requirements in the product development processes of complex systems and for fault-tolerant design and control has been already stated in Sect. 3.1. It is of paramount importance to consider requirements throughout the whole design process (compare [8, 29]). The main characteristics of sensible requirements are

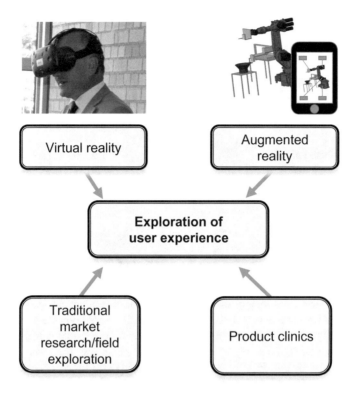

Fig. 4.11 Current means for user experience exploration

described in ISO-Standard 29148 [30]. Natural language or formalised natural language with restricted vocabulary and fixed sentence constructions can be used in order to document requirements. The objective of a requirement specification (including specifications and technical concept) is to formulate the requirements in such a way that a common understanding of the system to be developed is created. To achieve this understanding and to avoid ambiguity, certain rules should be followed. It is recommended to use short sentences and to avoid weak adjectives and adverbs (e.g. more beautiful).

4.3.1 Background

The efficient and effective development of complex systems requires a conscious management of requirements which is usually called Requirements Management (RM) [17, 28, 30]. RM includes, for example, the identification and definition of requirements, the documentation of requirements, the consensus building and the validation of requirements as well as measures for controlling and managing requirements (compare [17]). Numerous research activities concerning requirements

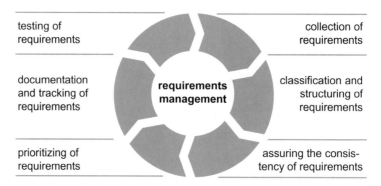

Fig. 4.12 A model of requirements management

management can be identified (compare [22, 32, 61]). It is possible to identify certain phases in requirements engineering (Fig. 4.12).

The phases are briefly explained in the next part of this sub-section:

- Collection of requirements: An initial step can be the accumulation of the essential requirements. In design methodology several checklists are available which can support this endeavor [24, 40, 42]. Other possible methods are benchmarking and "Quality Function Deployment" (QFD). The requirements concerning fault-tolerance can be found using techniques such as "Fault Tree Analysis" (FTA) [11] or "Failure Mode and Effects Analysis" (FMEA) [18]. In recent years it became clear that the accumulated collection cannot address all aspects of a planned technical system [3].
- Classification and structuring of requirements: For complex systems such as AGVs it is very difficult to retain a complete overview about the different requirements and the relationships between these requirements. Therefore, in literature a classification and structuring of requirements is frequently recommended [7, 17].
- Assuring the consistency of requirements: A set of requirements can only be consistent, if all competitive requirements are identified and if a compromise between these requirements is found. The phase "assuring the consistency of requirements" consequently also concerns the search for competitive requirements which are an important source for innovation and should be addressed early [8].
- Prioritizing requirements: In order to allow efficient and effective process planning and control, it is extremely important to arrange the requirements according to their priority [8].
- Documentation and tracking of requirements: Complex systems such as AGVs can dispose of several hundred important requirements. A structured and continuous recording, identifying and pursuing of these requirements in inevitable.
- Testing of requirements: In the closest sense, the testing is not a part of RM. However, only conceivable and logical, well managed requirements allow effective and efficient testing processes, carried out by means of simulations and physical tests.

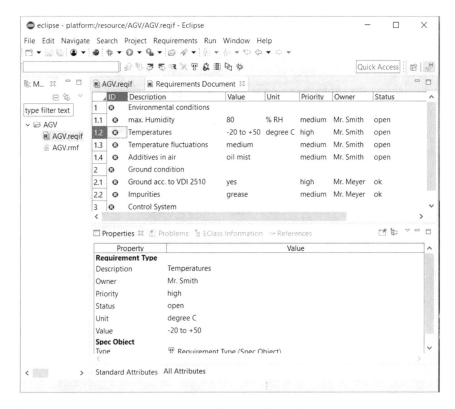

Fig. 4.13   Requirements management with eclipse "ProR"—AGV example

A large number of tools for requirements management is currently available on the market; an extensive overview is given by De Gea et al. [19]. It is also possible to use open source products, which can provide a full functionality such as the eclipse based tool "ProR". Figure 4.13 shows an example of AGV requirements modelled in ProR.

In Germany, a standard for the uniform exchange of requirements has been established, the so-called requirements interchange format (ReqIF), which is a format and data model that contains amongst others structures for requirements, their attributes, types, access rights and relations [23]. However, a complete connection to an existing product structure cannot yet be realised by means of these available tools and formats [28]. Therefore new approaches are needed in order to address this issue. One promising approach is using model based methods using graph-based languages.

4.3.2   Model Based Requirements Management

During the last two decades a framework was created which allows the realization of elaborate system models using graph-based languages [47] and was applied in

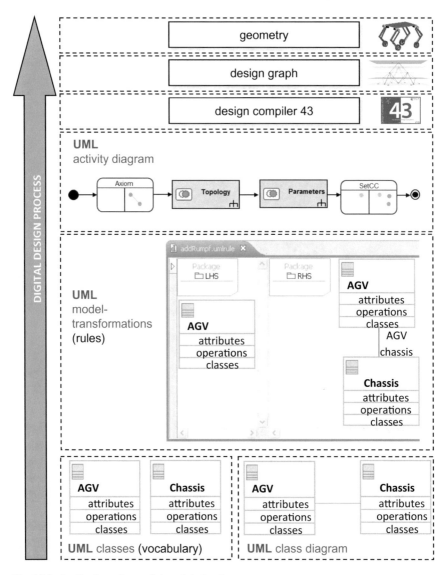

Fig. 4.14 Application of a graph-based design language

numerous industrial projects [4, 26, 28, 41, 44, 45, 59]. Figure 4.14 shows the general structure of the framework.

The basic elements of the framework are shown at the bottom of Fig. 4.14. The components of the sample product—an AGV—are shown as UML classes in the lower left corner. These components can be understood as the vocabulary of the design language. The connections between these classes are represented in the class diagram which shows the associations and the generalization/specialization inter-relationships between the classes as well as abstract classes which are not directly linked to certain elements of the AGV. The next level of Fig. 4.14 shows the model transformations (rules). These rules describe the transformation process of the structure of the AGV; they are specified by engineers. They are arranged in an UML activity diagram into a rule-sequence and can be executed in order to instantiate the classes for representing one specific product configuration. The execution process is carried out by the unique design compiler 43 [20] and leads to the design graph, which is can be understood as high-level central data model. From this model different interfaces can generate domain specific models, for instance geometry models or simulation models (CAD, FEM, etc.) in a completely automated manner. The content of the central data model can concern geometry but also physical loads, materials or functions. For an application of this kind of complex methodology it is necessary to analyse the application context in industry [50, 51].

4.3.3 Industrial Situation

The main research question in this sub-section can be formulated as: "*how can graph-based design languages be applied to requirements management in an advantageous manner?*" This kind of research is characterised by Blessing and Chakrabarti [12] as research project type five, which aims for an investigation of a novel approach. The general research procedure is sketched in Fig. 4.15.

The first stage called *research clarification* is applying an in-depth literature analysis in order to realise an initial problem understanding. On this basis a *descriptive study* is carried out which consists of analyses of the industrial situation. The results are a list of influencing factors which are reported in this subsection. The later stages *synthesis* and *analysis* can be carried out as an improvement cycle in order to allow an efficient development of the novel approach. Methods and tools are synthesised and continuously analysed with the concerned engineers.

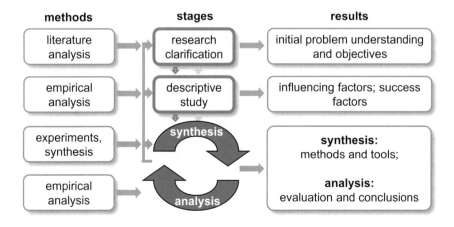

Fig. 4.15 General research procedure

The analysis of the industrial situation in the product development of AGVs firstly resulted in the insight that a distinct phase of the process is carried out in close contact with the costumer. Usually, AGVs have to carry out certain tasks within a complex system which is either already existing or is also under development. It is important to clarify the environmental conditions such as climatic conditions (temperatures, temperature fluctuations, humidity, etc.) and floor conditions (bearing capacity, contaminations, friction characteristic, etc.). For the frequent application as transporting AGV, the load transfer possibilities and necessities have to be clarified. A central point concerns the drive path and the presence or absence of obstacles and uphill or downhill sections. Last but not least, the possibilities and necessities of the control system need to be inquired as well as the desired levels of fault-tolerance.

The following points characterise the RM situation in industry:

- It is generally advantageous to elaborate a concise, cross-domain, integral requirements specification [3].
- Requirements are a central element in the product development process, because they provide indispensable information for product development engineers [54].
- End user requirements are vital for the success of development processes [36].
- The kano-model (compare Sect. 4.2) can clarify the divergences between different stakeholders [14].
- It is frequently advantageous to introduce virtual customers which represent a whole class of customer [22].
- Design requirements incorporate two different functions: to document the agreement what is wanted in the end product and to provide a basis from which the designer can proceed in synthesizing a solution [22].
- Unmanaged requirement change propagation can lead to product development project failure [39].
- Requirement models, which are integrated in graph-based domain and life-cycle spanning product models, may offer several practices for investigating the consistency of requirements [28].
- Requirement models, which are integrated in graph-based domain and product models, which stretch over the whole life-cycle, allow the tracking of requirements through all phases of the product life-cycle [28].

4.3.4 Application to the Requirements Management of Automated Guided Vehicles

The initial step of a product development process in graph-based design languages is the establishment of the vocabulary—the UML classes of the respective product (compare Fig. 4.14). For the development of an AGV some classes are explained in Fig. 4.16.

Each of the classes contains certain central attributes of the components such as dimensions (length, width and height) and mechanical properties such as the

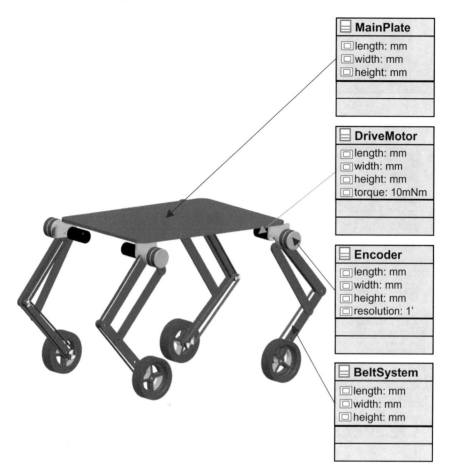

Fig. 4.16 Description of the components of an AGV using UML classes

maximum deliverable torque of a drive motor. Additionally, the class's operations can be documented in the lower part of the class—these operations could be certain calculations.

The connections of these classes are represented in the class diagram which shows the associations and the generalization/specialization interrelationships between the classes as well as abstract classes which are not directly linked to certain elements of the AGVs. The classes are connected to each other and organised in the class diagram. It is also possible to use classes from other class diagrams to extend the functionality of a certain design language. A sample class diagram is shown in Fig. 4.17.

One example for an abstract class is the class *MechElement* which includes attributes such as *elemCost* (i.e. the cost of the element in Euro) and *elemMass* (i.e. the mass of the element in grams) which all the concrete classes inherit from these abstract class. While the class diagram shows all possible types of combinations between the classes, the rules specify combinations of instances of the classes.

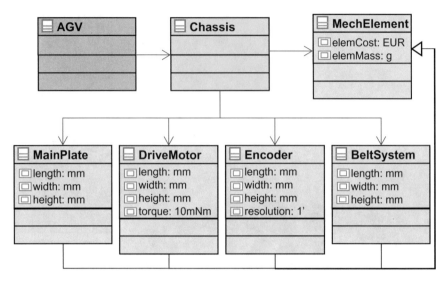

Fig. 4.17 Description of the structure of an AGV using an UML class diagram

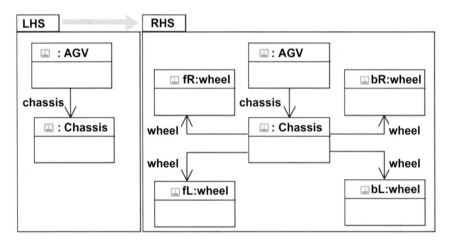

Fig. 4.18 Example for a graphical rule

For example, if it is determined in the class diagram that a AGV chassis has wheels, the rules specify which and how many. An example of a graphical rule is shown in Fig. 4.18.

The rules can be defined both graphically and via so-called "JavaRules" in which the UML transformations are generated by Java code. In general, any technical system can be abstracted to its parts and their abstract geometry. These parts can be organised by the class diagram. This vocabulary can be used together with rules in order to generate the geometrical model of the product. Figure 4.19 represents this digital development cycle.

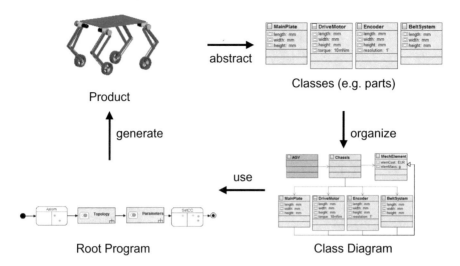

Fig. 4.19 Digital development cycle

The elements of the AGV are organised in the class diagram. In addition to this diagram, the engineers define rules which use and instantiate the class diagram. From this, the design graph can be generated—a central data model which allows to be used in multiple modelling and simulation computer programs such as "Computer Aided Design" (CAD), "Finite Element Method" (FEM) or "Multi Body Systems" (MBS).

This kind of digital product development process allows a unique integration of the requirements management. A central characteristic of this approach is the connection of requirements with a domain and life-cycle spanning product model. Requirements can be connected to modules or components of the AGV or to the AGV as a whole, if the requirements concerns desired characteristics that only the complete AGV can fulfil. That leads to an association between AGV classes and requirements. This could be realised by introducing an attribute *RelatedToPart* within the structuring of requirements. Figure 4.20 contains a class diagram of requirements of an AGV.

In the centre of Fig. 4.20 a typical requirement of an AGV is visible—the carrying capacity, i.e. how heavy the goods may be, which the AGV can transport. This requirement belongs to the functional requirements (see left side of Fig. 4.20) and is linked to the chassis of the AGV (see right side if Fig. 4.20). This connection to the product structure is also shown in Fig. 4.21.

On the top level, the complete AGV and its environment are located and are connected to requirements that concern specifically this level and cannot be assigned to distinct sub-assemblies or components. On the next level, the chassis as one main assembly is shown and is connected to requirements which can be specifically associated with this assembly. The next lower level contains two sub-assemblies—the front unit and the back unit—which both are connected to requirements that concern specifically these two units. On the sub-sub-assembly level a left arm unit and a right arm unit are visible (here both for the front unit)—these are connected to

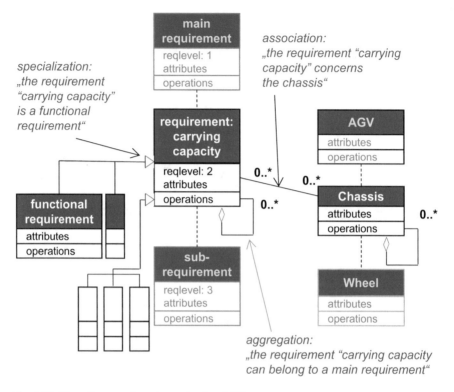

Fig. 4.20 Class diagram of requirements of an AGV

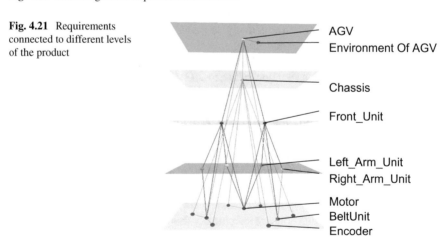

Fig. 4.21 Requirements connected to different levels of the product

certain requirements which concern these sub-sub-assemblies. On the lowest level of the product structure, components such as motor or encoder and sub-sub-sub-assemblies such as the belt unit are present and can be linked to requirements.

Fig. 4.22 Example for a design graph of instances of the product architecture and requirements

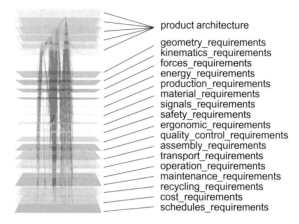

product architecture

geometry_requirements
kinematics_requirements
forces_requirements
energy_requirements
production_requirements
material_requirements
signals_requirements
safety_requirements
ergonomic_requirements
quality_control_requirements
assembly_requirements
transport_requirements
operation_requirements
maintenance_requirements
recycling_requirements
cost_requirements
schedules_requirements

In addition to the connection to the product structure (compare Fig. 4.20 on the left side), the system model can be associated to certain taxonomies, e.g. distinguishing functional requirements, which demand a functionality of the sub-assembly or component and quality requirements, which demand certain quality characteristics such as *corrosion free* as well as structural requirements, which directly demand certain structural characteristics or even the use of predefined modules (compare Fig. 4.20 on the right side). A more detailed taxonomy is shown in Fig. 4.22.

The upper levels in Fig. 4.22 represent the product architecture as shown in Fig. 4.21. On the lower levels, the requirements are structured with regard to the taxonomy of Becattini et al. [7] (Geometry-, Kinematics-, Forces-, Energy-, Production-, Material-, Signals-, Safety-, Ergonomic-, Quality control-, Assembly-, Transport-, Operation-, Maintenance, Recycling-. Cost- and Schedules requirement) which is mostly utilised in check-lists for conceptual design.

The main advantages of the model based approach in the certain phases of RM (compare Fig. 4.12) and the connection to the aspects of fault-tolerance are explained in the next part of this sub-section:

- Collection of requirements: The collection of requirements is a crucial step in industrial product development [7]. The connection of requirements to the product structure can easily lead to blind spots in the collection of requirements. It would be very unusual, if a subassembly or component is not associated with a single requirement. Such blind spots allow a goal-directed search for further important requirements. It is also possible to represent functions in the central model (compare [45]). This fact also supports the detection of blind spots in the collection of requirements. As it is possible to model fault-tolerance on the functional level (compare Sect. 3.2), also the collection of requirements concerned with fault tolerance is fostered.
- Classification and structuring of requirements: An enhanced classification and structuring of requirements is the main emphasis of model based requirements management. The nearly unlimited possibilities for associations with the product structure, which also contains domain spanning information, and for integrating

a multitude of taxonomies allow a very detailed classification and structuring of requirements and establish a poly-hierarchical model (compare [28]). This presents an optimum basis for the numerous synthesis and analysis activities concerning requirements.

- Consistency of requirements: Model based requirements management offer various possibilities to assure the consistency of requirements. The associations to different taxonomies and to the product structure expand the information content greatly and allow the application of consistency checking algorithms [28]. The development of such algorithms is ongoing.
- Prioritizing requirements: Engineering managers in industry underline the enormous importance of activities concerned with prioritizing requirements [8]. A profound prioritization requires domain spanning explorations of the complete design space. This can be fostered by executable design languages and wide possibilities to vary configurations in terms of parameters, topology and solution domain. The transparency resulting from model based requirements management can support the prioritization of requirements concerning fault-tolerance, because sometimes these requirements are competing with economic requirements and the proof of their importance is sometime not trivial.
- Documentation and tracking of requirements: Model based requirements management with graph-based design languages can bring the structure of requirements to the next organizational level by integration in a domain and life-cycle spanning product model. A continuous tracking is supported by unique identifiers and a central data model. The documentation and tracking of requirements which concern fault tolerance is extremely important and central, because these requirements can influence many sub-assemblies and components such as sensors, actuators and control systems. Model based requirements management offers the possibility to enhance the visibility of these central requirements.
- Testing of requirements: The connection of requirements to the product structure enhances the overview of the numerous tests in different domains on several system integration levels. Several simulations can be carried out early in a large solution space thus reducing the need for extensive testing in late stages. An exhaustive system model especially fosters the testing of requirements concerning fault-tolerance, because the propagation of faults is transparent in such models.

The structured documentation of requirements in the central data model presents an optimum basis for the domain-spanning system design.

4.4 System Design

Once the requirements have been initially established, structured and documented, the development of a principle solution which consists of the central physical and logical operating concept can be started in the system design stage. Major challenges in this stage are the interdisciplinary nature of the operating concept and the fact that no common abstract functional language is available up to today. Certain language haves the potential to be used, such as the "System Modelling Language" (SysML) [53, 60] but have not yet been accepted throughout the industry and the models are not yet understood by traditional engineers in the mechanical and electrical domain. Additionally, SysML has the disadvantage that the models cannot be instantiated and compiled, therefore SysML is generally limited to be used in the abstract stages in system design and does only allow configurations exploration to a very limited degree. System models in UML are also not simply understood by older engineers. However, they exhibit the advantage that they can be combined with graph-based design languages and can be instantiated and compiled (machine executed) and thus allow a wide range of configuration exploration (compare Sect. 3.1). It is important to point out that this exploration is not limited to configurations which concern the change of parameters but also topological changes and domain changes. Therefore, early explorations of a large, interdisciplinary solution space become feasible. In [52] the following aspects of information have been found to be necessary for an abstract interdisciplinary system model:

- Input/Output: An abstract system model should be able to model the flow of matter, energy and signals through the technical system under development at least in terms of input and output states.
- Aggregation/Decomposition: An abstract system model should allow to represent the hierarchical structure of the technical system under development.
- Function: An abstract system model should allow to represent the functional relationships of the technical system under development.
- Generalization/Specialization: An abstract system model should dispose of the capability to represent the fact that certain elements will belong to a certain class, e.g. that components of the suspension and components of the drive train are both "mechanical components" which share certain attributes such as dimensions or weight.

Ramsaier et al. [44, 45] have been able to show these aspects in model-based engineering with graph-based languages on the example of the development of a Quadrocopter. A common system model can allow to combine important structural and functional aspects of all involved disciplines in one consistent model and to explore the design space including topological and domain changes for sub-systems and components as well as to achieve a logically and functionally optimum principle solution concept. In this stage, the abstract fault-tolerance mechanisms can be defined in the common system model. The resulting concept can serve as a solid basis for the domain specific design.

4.5 Domain Specific Design

Main challenges in the domain specific design are the complexity of current products such as AGVs and the multitude of requirements from all life phases of the technical system under development. Fortunately, in all domains powerful methodologies and tools have been developed in order to assist this stages. Improvements in this stage can be achieved, if an ongoing requirements management and an updated domain-spanning system model are present and available for the involved engineers. In the different domains the engineers have to realise the desired level of fault-tolerance, which is defined in the requirements, by combining concepts such as redundancy, robust design, prognosis and virtual sensors. It is desirable to clearly assign the responsibility to control and report the presence or absence of fault-tolerance to one involved engineer, in order to prevent later changes if deficiencies in fault-tolerance are detected during the verification and validation of the concept (compare Sect. 4.7).

4.6 System Integration

The system integration stage is characterised by integration endeavours on many stages, generally preparing the system for certain verification/validation steps on different levels. This can be explained on the example of an AGV. It would not be sensible if one would first try to design every detail of an AGV and then to test the whole product, if it will perform the desired tasks. The drawback of this kind of procedure is that design flaws or less then optimum solutions would be detected very late in the process causing enormous expenditures. Therefore, it is sensible to start the verification/validation on the component level, making sure that components like electrical motors are working correctly (for supplier products this has already been done before the project however not all working conditions might have been considered). The next integration level encompasses the sub-assemblies such as the suspension arm equipped with a drive motor and an encoder. These sub-assemblies dispose of a functionality which can already be tested without assembling a complete AGV. It is important to note that calculation and simulation also play a really important role. The final integration level must also include a profound testing of the communication between the sub-assemblies. The interfaces between sub-assemblies are of prominent importance [52]. The interfaces can be described incorporating:

- aspects of matter, energy or signals flows through the respective interface,
- functional, geometrical and syntactical compatibility of both sides of the respective interface and
- common functions which require coordinated characteristics of both sides of the respective interface (e.g. the sealing of an electrical connector and the appropriate geometry and surface and the other side).

As stated above, the system integration stage is linked to the verification and validation on many levels.

4.7 Verification and Validation

The *verification* of a technical system verifies that this system, in its development, conforms to the specified requirements, whereas the *validation* needs to rely on broader field tests which evaluate whether specified customer needs are fulfilled. For the *verification*, calculation and simulation are tools which can enhance its efficiency. Additionally, so-called "X in the loop" verification procedures (e.g. hardware in the loop) combine physical tests with simulations and foster early verification which can greatly reduce the development risk. Domain spanning system models which dispose of a modular structure can ease and accelerate the application of "X in the loop" verification procedures. An ideal situation is created, if also the requirements which are specified for the respective technical systems are present in the same system model. In general, the *verification* of fault-tolerance is a straight forward process within which the fault is realised in an artificial environment (which is similar enough to the realistic environment) and the reaction and behaviour of the respective component, sub-assembly or system is recorded and evaluated. For instance, for an AGV which should be fault-tolerant with regard to the fault "loss of tire pressure" the engineers would reduce the pressure in one or more tires, would record the driving behaviour in an artificial but realistic operation scenario, would compare this behaviour to the specified requirements and would verify this specific aspect of the fault-tolerance. The *validation* of the fault-tolerance requires field tests with real costumers—in such scenarios the incitement of a fault might be problematic, because the equipment of the costumer and even the health of the employees might be at risk. Consequently, the validation of fault-tolerance requires conscious planning by the responsible engineers with deep considerations of worst case scenarios during the validation. In some cases the validation of fault-tolerance with regards to certain faults might even be impossible; in this case the engineers need to rely on verification results together with the extensive application of FTA [11] and FMEA [18].

It has to be pointed out that an early knowledge about the fault-tolerant qualities of a technical systems can be extremely helpful. The general insight that product properties should by determined early in product development processes (compare [9, 37]) is also valid for the fault-tolerant qualities, because the costs of changes necessary in order to achieve a certain intended level of fault-tolerance rise progressively during the product life cycle (Fig. 4.23).

In Fig. 4.23, the modification costs necessary in order to achieve an intended level of fault-tolerance are shown (they increase progressively) together with the possibilities to modify. These possibilities are getting smaller during the life-cycle, because certain things such as the basic concept and components which are produced in complicated tools normally cannot be changed in late phases. Additionally, the knowledge level of fault-tolerance (i.e. the knowledge of the engineers that the technical system will definitely achieve certain fault-tolerant qualities and functionalities) is shown for two cases: The conventional case is the one, when no specific efforts are directed towards an early knowledge of the fault-tolerant qualities and functionalities of the technical system under development. The improved case is the one, when engineers

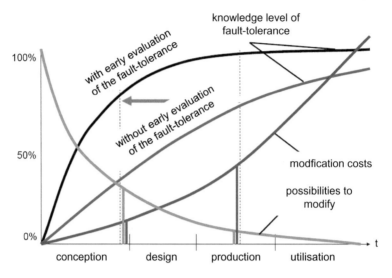

Fig. 4.23 Importance of an early evaluation of the fault-tolerance of a technical system

consciously try to explore the fault-tolerant qualities and functionalities of the technical system under development early in the process. The corresponding approaches can be summarised under the notion "early evaluation of fault-tolerance". The "early evaluation of fault-tolerance" can be understood as a form of front-loading and can include methods such as "Fault Tree Analysis" (FTA) and "Failure Modes and Effects Analysis" (FMEA), but most prominently consciously planned experiments and simulations which explore the behaviour of the technical system in the case of probable faults. Further research is needed in order to collect and systematise appropriate guidelines, methods and tools for a goal-directed early evaluation of fault-tolerance.

4.8 Summary

Complex technical systems such as automated guided vehicles realise their functionality through an interplay of mechanical, electrical and electronic components and sub-assemblies as well as software. The product development process of such systems is challenging and can only be carried out in an effective and efficient manner, if elaborate processes and integrated models are available. This chapter has presented central aspects of such processes. On the example of requirements engineering, which is an essential element in these product development processes, the possibilities of system modelling with graph-based languages has been explained in detail. Throughout the process, aspects of fault-tolerant control and fault-tolerant design play a prominent role.

References

1. Albers, A., Brudniok, S.: Methodische Entwicklung von hochintegrierten mechatronischen Systemen am Beispiel eines humanoiden Roboters. In: VDI-Bericht 1971: Mechatronik 2007 -Innovative Produktentwicklung -Maschinenbau, Elektrotechnik, Informationstechnik (2007)
2. Albers, A., Deigendesch, T., Meboldt, M.: Information integration and cooperation in product development by utilization of wikis. In: Proceedings of the Conference Tools and Methods for Competitive Engineering, TMCE 2008 (2008)
3. Almefelt, L., Berglund, F., Nilsson, P., Malmqvist, J.: Requirements management in practice: findings from an empirical study in the automotive industry. Res. Eng. Des. **17**(3), 113–134 (2006)
4. Arnold, P., Rudolph, S.: Bridging the gap between product design and product manufacturing by means of graph-based design languages. In: Proceedings of the 9th International Symposium on Tools and Methods of Competitive Engineering (TMCE 2012) (2012)
5. Bathelt, J., Joensson, A., Bacs, C., Kunz, A., Meier, M.: Conceptual design approach for mechatronic systems controlled by a programmable logic controller (PLC). In: Proceedings of ICED03, International Conference on Engineering Design (2003)
6. Bathelt, J., Joensson, A., Bacs, C., Dierssen, A., Meier, M.: Applying the new VDI design guideline 2206 on mechatronic systems controlled by a PLC. In: Proceedings of ICED05, International Conference on Engineering Design (2005)
7. Becattini, N., Cascini, G., Rotini, F.: Requirements checklists: benchmarking the comprehensiveness of the design specification. In: Proceedings of the 20th International Conference on Engineering Design (ICED15)
8. Bernard, R., Irlinger, R.: About watches and cars: winning R and D strategies in two branches. In: International Symposium "Engineering Design The Art of Building Networks" (2016)
9. Bernard, R., Stetter, R.: Early determination of product properties. In: Proceedings of International Conference on Engineering Design ICED 97, Tampere, vol. 2, pp. 675–680 (1997)
10. Bernardi, M., Bley, H., Schmitt, B.: Integrating a mechatronics-oriented development process into a development department. In: Proceedings of CIRP, Budapest (2004)
11. Bertsche, B.: Reliability in Automotive and Mechanical Engineering. Springer, Berlin (2008)
12. Blessing, L.T.M., Chakrabarti, A.: DRM, a Design Research Methodology. Springer, Berlin (2009)
13. Bludau, C., Welp, E.: Semantic web services for the knowledge-based design of mechatronic systems. In: Proceedings of ICED05, International Conference on Engineering Design (2005)
14. Borgianni, Y., Rotini, F.: Stakeholders diverging perceptions of product requirements: implications in the design practice. In: Proceedings of the 20th International Conference on Engineering Design (ICED15) (2015)
15. Braun, S., Lindemann, U.: Multiplanare vernetzungen abhngigkeiten zwischen produktkonzept, produkterstellungsprozess und ressourcenverbrauch in der mechatronik. In: Multiplanare Vernetzungen Abhngigkeiten zwischen Produktkonzept, Produkterstellungsprozess und Ressourcenverbrauch in der Mechatronik (2007)
16. Breuer, H., Wogatzky, M., Steinhoff, F.: User clinic formats and their value contribution to innovation projects. In: Proceeding of the 2nd International Society for Professional Innovation Management (ISPIM) Innovation Symposium, New York (2009)
17. Buehne, S., Herrmann, A.: Handbuch Requirements Management nach IREB Standard. Aus- und Weiterbildung zum IREB Certified Professional for Requirements Engineering Advanced Level Requirements Management. IREB e.V. (2015)
18. Carlson, C.S.: Effective FMEAs: Achieving Safe, Reliable, and Economical Products and Processes Using Failure Mode and Effects Analysis. Wiley, New York (2012)
19. Carrillo de Gea, J.M., Nicolas, J., Fernandez Aleman, J.L., Toval, A., Ebert, C., Vizca, A.: Requirements engineering tools: capabilities, survey and assessment. Inf. Softw. Technol. **54**(10), 1142–1157 (2012)
20. Compiler 43: developed by the IILS mbH in cooperation with the University of Stuttgart

21. Daenzer, W.F., Huber, F.: Systems Engineering Methodik und Praxis. Verlag Industrielle Organisation, Zurich (2002)
22. Darlington, M.J., Culley, S.J.: A model of factors influencing the design requirement. Des. Stud. **25**(4), 329–350 (2004)
23. Ebert, C., Jastram, M.: ReqIF: seamless requirements interchange format between business partners. IEEE Softw. **29**(5), 82–87 (2012)
24. Ehrlenspiel, K., Meerkamm, H.: Integrierte Produktentwicklung. Denkabläufe, Methodeneinsatz, Zusammenarbeit. Carl Hanser Verlag, Munich (2013)
25. Gausemeier, J., Moehringer, S.: New guideline VDI 2206 a flexible procedure model for the design of mechatronic systems. In: Proceedings of the 14th International Conference on Engineering Design (ICED03) (2003)
26. Gross, J., Rudolph, S.: Generating simulation models from UML a FireSat example. In: Proceedings of the 2012 Symposium on Theory of Modeling and Simulation DEVS Integrative M and S Symposium (2012)
27. Hellenbrand, D.: Transdisziplinre Planung und Synchronisation mechatronischer Produktentwicklungsprozesse. Dissertation Technische Universitaet Muenchen (2013)
28. Holder, K., Zech, A., Ramsaier, M., Stetter, R., Niedermeier, H.-P., Rudolph, S., Till, M.: Model-based requirements management in gear systems design based on graph-based design languages. Appl. Sci. **7** (2017)
29. Hruschka, P.: Business Analysis und Requirements Engineering: Produkte und Prozesse nachhaltig verbessern. Hanser, Munich (2014)
30. ISO/IEC/IEEE 29148:2011: systems and software engineering - life cycle processes - requirements engineering
31. Jansen, S., Welp, E.: Model-based design of actuation concepts: a support for domain allocation in mechatronics. In: Proceedings of ICED05, International Conference on Engineering Design (2005)
32. Jiao, J., Chen, C.H.: Customer requirement management in product development: a review of research issues. Concurr. Eng. Res. Appl. **14**(3), 173–185 (2006)
33. Kano, N., Seraku, N., Takahashi, F., Tsuji, S.: Attractive quality and must-be quality. J. Jpn. Soc. Qual. Control **14**(2), 147–156 (1984)
34. Klein, T.P.: Agiles Engineering im Maschinen- und Anlagenbau. Dissertation Technische Universitaet Muenchen (2016)
35. Klein, T.P., Reinhart, G.: Towards agile engineering of mechatronic systems in machinery and plant construction. Procedia CIRP **52**, 68–73 (2016)
36. Li, H.-X., Lu, X.: System Design and Control Integration for Advanced Manufacturing. Wiley, Zurich (2015)
37. Lindemann, U., Stetter, R.: Industrial application of the method "early determination of product properties". In: Proceedings of the ASME 1998 Design Engineering Technical Conferences and Computers in Engineering Conference, Atlanta (1998)
38. Maximini, D.: Scrum - Einfhrung in der Unternehmenspraxis. Springer, Berlin (2013)
39. Morkos, B., Mathieson, J., Summers, J.D.: Comparative analysis of requirements change prediction models: manual, linguistic, and neural network. Res. Eng. Des. **25** (2014)
40. Pahl, G., Beitz, W., Feldhusen, J., Grote, K.H.: Engineering Design: a Systematic Approach. Springer, Berlin (2007)
41. Pfitzer, S., Rudolph, S.: Re-engineering exterior design: generation of cars by means of a formal graph-based engineering design language. In: Proceedings of the 16th International Conference on Engineering Design (ICED07) (2007)
42. Ponn, J., Lindemann, U.: Konzeptentwicklung und Gestaltung technischer Produkte. Springer, Berlin (2011)
43. Preussig, J.: Agiles Projektmanagement. Haufe (2015)
44. Ramsaier, M., Spindler, C., Stetter, R., Rudolph, S., Till, M.: Digital representation in multicopter design along the product life-cycle. Procedia CIRP **62**, 559–564 (2016)

45. Ramsaier, M., Holder, K., Zech, A., Stetter, R., Rudolph, S., Till, M.: Digital representation of product functions in multicopter design. In: Proceedings of the 21st International Conference on Engineering Design (ICED 17) vol 1: Resource Sensitive Design, Design Research Applications and Case Studies (2017)

46. Roelofsen, J.: Situationsspezifische Planung von Produktentwicklungsprozessen. Dissertation Technische Universität München (2011)

47. Rudolph, S.: Aufbau und einsatz von entwurfssprachen fr den ingenieurentwurf. In: Forum Knowledge Based Engineering, CAT-PRO (2003)

48. Schwaber, K.: Agile Project Management with Scrum. Microsoft Press, Redmond (2004)

49. Stetter, R.: Method implementation in integrated product development. Dissertation Technische Universitaet Muenchen. Dr.-Hut (2000)

50. Stetter, R.: Adoption and refusal of design strategies, methods, and tools in automotive industry. In: Chakrabarti, A., Lindemann, U. (eds.) Impact of Design Research on Industrial Practice. Tools, Technology, and Training, pp. 451–464. Springer, Cham (2015)

51. Stetter, R., Lindemann, U.: Transferring methods to industry. In: Clarkson, P.J., Eckert, C.M. (eds.) Design Process Improvement. Springer, Berlin (2005)

52. Stetter, R., Paczynski, A., Zajac, M.: Methodical development of innovative robot drives. J. Mech. Eng. (Strojniski vestnik) **54**, 486–498 (2008)

53. Stetter, R., Seemüller, H., Chami, M., Voos, H.: Interdisciplinary system model for agent-supported mechatronic design. In: Proceedings of the 18th International Conference on Engineering Design (ICED11)

54. Sudin, M.N., Ahmed-Kristensen, S., Andreasen, M.M.: The role of a specification in the design process: a case study. In: Proccedings of the International Design Conference - Design 2010 (2015)

55. Ullah, S., Tamaki, J.: Analysis of Kano-model-based customer needs for product development. Syst. Eng. **14**(2), 154–172 (2011)

56. VDI 2206:2004-06: design methodology for mechatronic systems

57. VDI 2221:1993-05: systematic approach to the development and design of technical systems and products

58. VDI 2422:1994-02: systematical development of devices controlled by microelectronics

59. Vogel, S., Arnold, P.: Object-orientation in graph-based design grammars (2017). arXiv:1712.07204

60. Weilkiens, T.: Systems Engineering mit SysML/UML: Anforderungen, Analyse, Architektur. Dpunkt.verlag, Heidelberg (2014)

61. Zhang, Z., Li, X., Liz, Z.: A closed-loop based framework for design requirement management. In: Moving Integrated Product Development to Service Clouds in the Global Economy. Proceedings of the 21st ISPE Inc. International Conference on Concurrent Engineering (2014)

Chapter 5
Design of Virtual Diagnostic Sensors for an Automated Guided Vehicle

Diagnostic sensors, i.e. sensors that allow "Fault Detection and Identification" (FDI), are of paramount importance for fault-tolerant control. This chapter describes an innovative approach aiming at synthesising sensor information in cases where no sensor data are available or the available sensor data are not reliable enough. In such cases, mathematical models can still allow the generation of reliable sensor information, thus creating virtual sensors. The development and application of an innovative virtual sensor is explained following the example of one automated guided vehicle (AGV). In general, the fault-tolerance of systems which use AGVs for logistic tasks is rather high, because the breakdown of one AGV can often be compensated. However, certain conditions such as slippery floors and tire air pressure losses can be problematic for the AGV operation. Diagnostic sensors are one cornerstone for more fault-tolerant AGVs and are in the centre of this chapter. The results presented in this section are based on the publications [14, 15].

The explanation of this approach starts with a conclusion of the state of the art (Sect. 5.1). The specific research questions are addressed in Sect. 5.2. Section 5.3 introduces essential preliminary considerations, which are necessary to approach the investigated problem and discusses the description of the discrete event system. The unique design and realisation of the automated guided vehicle, which is used for the sake of explanation and validation, is explained in Sect. 5.4. Section 5.5 describes the mathematical foundation. In Sect. 5.6 a strategy for the design of the virtual diagnostic sensor is proposed. Section 5.7 presents the outcome of the application of the proposed strategy to the AGV. In Sect. 5.8 experimental results are shown and explained, which clearly display the performance of the developed approach.

© Springer Nature Switzerland AG 2020   93
R. Stetter, *Fault-Tolerant Design and Control of Automated Vehicles and Processes*, Studies in Systems, Decision and Control 201,
https://doi.org/10.1007/978-3-030-12846-3_5

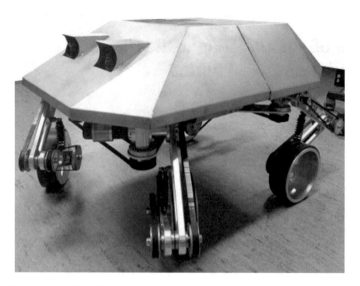

Fig. 5.1 Automated guided vehicle

5.1 State of the Art

AGVs are currently used in many areas of human life such as industrial production and transportation. However, the fault-tolerant control of these vehicles still provides a challenge for companies and institutions. One central challenge is the development and implementation of reliable sensors. Sensors are needed in order to measure aspects of the (internal) states of an AGV as well as aspects of the (external) states of its surrounding, maybe even including other AGVs. In certain cases, sensors are even required for enabling the main functionalities of AGVs. One prominent example is an innovative AGV developed at the Hochschule Ravensburg–Weingarten (compare Fig. 5.1).

This AGV disposes of a unique steering system which allows unlimited manoeuvring possibilities but requires reliable sensor data for its operation. The unique steering system is sketched in Fig. 5.2.

Four wheels are individually driven by controlled electrical motors. These wheels are fixed on arms which can rotate around an axis which is not located in the centre of the contact of the wheel with the ground. Consequently, the longitudinal forces of the wheels which are necessary to drive the AGV also result in moments with regard to this axis which allow to steer the direction of the arm and the connected wheel. For increased controllability and driving performance, in the AGV the two front arms and the two back arms are connected by belts thus achieving the same steering angle at both wheels. This allows strong accelerations and diminishes the influence of slippery surfaces on one of the wheels.

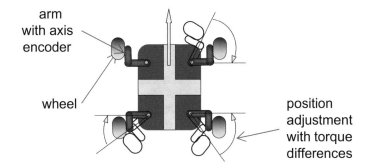

Fig. 5.2 Steering system

As stated above, this steering system requires reliable sensor data, which can be achieved by a supplementary application of virtual sensors. Virtual sensors employ appropriate mathematical models and additional sources of information such as other sensors or states of actuators in order to synthesise virtual measurements. These virtual sensors can enhance the fault-tolerance of AGVs. The main objectives of the scientific activity described in this section are the design of virtual sensors which can determine the forces and the torques that are acting onto an AGV. For this purpose, a quadratic boundedness approach is applied. This approach allows to include bounded disturbances acting onto the AGV. This approach does not rely on the application of complex tire models. Instead, measurements of acceleration and yaw rate sensors are used in order to synthesise reliable virtual information concerning forces and torques. These measurements enable several diagnostic processes such as the detection or prevention of faults.

The main field of application of AGVs are scenarios with prominent material flows. AGVs are used in assembly lines, warehouses and production plants [12] and exhibit several advantages compared to other kinds of logistics systems. AGVs are more flexible, substitutable and intelligent than other systems, they occupy a smaller floor space and require relatively small amounts of time and cost for their initial installation [18]. Unfortunately, the immense potential of AGVs in terms of efficiency and flexibility is not yet exploited, because frequently AGVs are only used for simple assignments such as the loading and unloading of goods and only fixed guiding technologies are applied, for instance magnetic guidance or optical guidance [18]. One main problem causing these situations can be the hardware design [18]. AGVs with conventional steering systems, for instance classical Ackermann steering, exhibit limited manoeuvring capabilities and may waste precious space for their movement. Additional main problems causing the rather low application ratio of AGVs are the high cost of sensors and the complexity and difficulty of sensor data filtering, sensor data plausibility assessment and sensor fusion. If AGVs are operating in real environments and executing real tasks, some parameters, for instance slip, exact wheel diameter and mass, are altered during the operation, because of uneven load distribution, manufacturing imperfections and unequal loading of goods [10]. It is difficult to gather reliable sensor data under such conditions. Also, the relatively low application ratio can be caused by the less than sufficient flexibility, availability and reliability of current AGV designs. The central objective of the research presented in this chapter is to develop promising approaches for an optimised design of virtual sensors. Two approaches are combined in order to contribute to a higher efficiency and reliability of AGVs: the innovative design of an AGV, which is presented in this chapter, allows unlimited capabilities for manoeuvring and the virtual diagnostic sensor design provides reliable information from sensors without causing additional expenditures.

5.2 Research Question and Structure

One of the main advantages of fault-tolerant control systems is that they can meet control objectives, even if one or more fault(s) occur. However, frequently the application of these promising systems is prevented because of high sensor cost or a lack of reliable sensors [4, 8, 16]. A promising approach to tackle this problem is the application of virtual sensors, which can be used for diagnostic purposes. Virtual sensors use mathematical models of the process and other measurements, which are available on and around an AGV, in order to estimate certain unmeasured variables [19]. Several research approaches aim at the design of virtual diagnostic sensors: for instance observer-based [2, 7], Kalman filter-based [5, 9] and parameter identification-based [3] approaches.

The research described in this chapter is focused on an innovative approach for the design of virtual sensors which gather measurements of longitudinal forces and torques acting onto an AGV. In this innovative approach a quadratic boundedness (QB) approach [1] is used. This approach enables the inclusion of bounded disturbances acting onto the AGV. Additionally, the approach prevents unnecessary state estimation, because information concerning the state is completely available by means of measurements. It can be perceived as one central advantage of this approach that the application of sophisticated tire models is not necessary. Frequently, the necessity to rely on this kind of models restricts the performance of the current approaches presented in literature [6, 11].

Thus, the main research question can be formulated: "*how can reliable virtual sensors which are based on a quadratic boundedness approach and intended to provide measurements of longitudinal forces and torques of automated guided vehicles be designed?*"

This approach is explained and validated using an innovative design of an AGV, which allows unlimited manoeuvring capabilities but requires reliable sensor information. This system is described as discrete time system.

5.3 Description of Discrete Time Systems

Firstly, the discrete-time system is described using the following equation:

$$x_{k+1} = Ax_k + Bu_k + Bd_k + Wwk. \tag{5.1}$$

In this equation, $x_k \in \mathbb{X} \subset \mathbb{R}^n$ is the state vector, $u_k \in \mathbb{U} \subset \mathbb{R}^r$ stands for the input, $d_k \in \mathbb{R}^{n_d}$ is an unknown input and $wk \in \mathcal{E} \subset \mathbb{R}^{n_w}$ is an exogenous disturbance vector. $V_k = x_k^T P x_k$ describes a Lyaponov candidate function, i.e. a scalar function applied for proving the stability of an equilibrium. For an unforced $u_k = 0$ and unknown input-free $d_k = 0$ system (5.1), the subsequent definitions are appropriate:

The system (5.1) is strictly quadratically bounded for all allowable $wk \in \mathcal{E}, k \geq 0$, if

$$V_k > 1 \Rightarrow V_{k+1} - V_k < 0, \tag{5.2}$$

for any $wk \in \mathcal{E}$.

A set $\mathcal{E}_x$ is a robust invariant set for the system (5.1) for all allowable $wk \in \mathcal{E}$ if

$$x_k \in \mathcal{E}_x \Rightarrow x_{k+1} \in \mathcal{E}_x, \tag{5.3}$$

for any $wk \in \mathcal{E}$.

It is important to point out that the strict quadratic boundedness allows to decrease the value of the Lyapunov function V_k, i.e, it means that $V_{k+1} < V_k$ for any $wk \in \mathcal{E}$ when $V_k > 1$. If (5.1) is quadratically bounded and if at least one vector $Wwk \neq 0$ exists , then this quadratic boundedness is always strict [1]. Additionally, the notation of the quadratic boundedness can be articulated applying the theory of invariant sets [1].

The set $\mathcal{E}_x = \{x : x^T P x \leq 1\}$ is an invariant set for any $wk \in \mathcal{E}$, in the case that $x_k \in \mathcal{E}_x$ implies $x_{k+1} \in \mathcal{E}_x$. Consequently, if $V_k > 1$ then x_k is outside an invariant set, and therefore, $V_{k+1} < V_k$. This means that V_k is getting larger until x_k is outside $\mathcal{E}_x$.

This description can be applied for realising a virtual sensor for an automated guided vehicle.

5.4 Design and Realisation of an Automated Guided Vehicle

The central design goals during the development of the innovative AGV at the Hochschule Ravensburg-Weingarten were the synthesis of unlimited manoeuvring possibilities with a rather simple mechanical design and a high fault-tolerance (Fig. 5.1). Earlier work at the same university resulted in the design and the realisation of a production platform with a patented steering principle [20]. This production platform excelled with very good manoeuvring possibilities [13], but necessitated eight individual drive motors and could only operate on rather flat floors. The innovative design, which is described in this chapter, enables driving in uneven environments and only needs four drive motors. The AGVs main frame consists of four arms which each dispose of a drive motor and a spring/damper system. Generally, all arms can freely rotate around a defined axle outside the centre of the respective wheel (Fig. 5.2). However, the front arms and the back arms are connected by belts thus reducing the degrees of freedom and allowing better dynamical controllability and stronger accelerations. Each drive motor unit also includes an angle encoder. This enables the determination of the angle, the angular velocity and the angular acceleration of each wheel. A further set of two angular encoders can measure the steering angle of both the front wheels and the back wheels. The speed and torque of the four drive motors can be controlled individually, because four individual motor controllers (electronic position control EPOS—communicating via the CANopen protocol) are used. The ability to control the torque of four independent drive motors and a unique geometrical layout leads to a steering system which is using the torque differences between wheels to steer the axles of a vehicle. As a consequence of the application of this steering system, the AGV is able to drive directly in any desired direction without time- and space-consuming turning manoeuvres and can turn around its own centre. Especially in narrow spaces, which are common in current production environments, this characteristic is extremely beneficial. It is another advantage that the mechanical design is comparatively simple, because no dedicated steering motors are required. This quality causes a high robustness. In addition to the six angular sensors at the wheels and the steering axes, the robot is equipped with sensors which determine the global acceleration, the global velocity and the global yaw rate. The resulting sensor data is available via a Bluetooth connection. In Fig. 5.3 the important parameters of the steering system of the AGV together with all the considered forces and the respective parameters are shown; they are additionally listed in Table 5.1.

Each arm, which holds a wheel, disposes of a spring suspension with damping functionality which is intended to absorb impacts from uneven terrain (Fig. 5.4).

Two different modes can be selected for controlling the movements of the AGV: the "manual driving mode" and the "autonomous driving mode". A central control unit calculates the necessary steering angles for the front and the back wheels and sends appropriate angular velocity commands to the four drive motors. These commands allow to achieve the desired steering angle, the desired driving direction, the desired Instantaneous Centre of Rotation (ICR), which is often infinitely far away, and the desired vehicle velocity.

5.5   Mathematical Model of the Automated Guided Vehicle

From Fig. 5.3, the conclusion can be deduced that the force which is causing longitudinal motion is given by the subsequent equation:

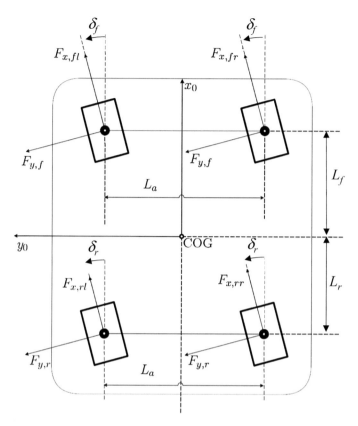

Fig. 5.3 Steering system

Table 5.1 AGV parameters

Variable	Unit	Value
m	kg	50
L_f	m	0.25
L_r	m	0.25
I_z	kg/m^2	89.18
C_r	kg/$^\circ$	1.86
C_f	kg/$^\circ$	1.86
L_a	m	0.16
R_e	m	0.09
I_{xw}	kg/m^3	0.00209089

Fig. 5.4 Suspension system

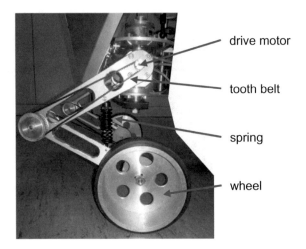

drive motor

tooth belt

spring

wheel

$$F_x = \cos(\delta_f)(F_{x,fl} + F_{x,fr}) + \cos(\delta_r)(F_{x,rl} + F_{x,rr}) +$$
$$- \sin(\delta_f)F_{y,f} - \sin(\delta_r)F_{y,r}. \tag{5.4}$$

Similarly, the lateral forces can be analysed using the subsequent equation:

$$F_y = \sin(\delta_f)(F_{x,fl} + F_{x,fr}) + \sin(\delta_r)(F_{x,rl} + F_{x,rr}) +$$
$$+ \cos(\delta_f)F_{y,f} + \cos(\delta_r)F_{y,r}. \tag{5.5}$$

In this equation, the longitudinal wheel forces obey:

$$I_{xw}\dot{\omega}_{fl} = p_{f,l}T - F_{x,fl}R_e, \tag{5.6}$$
$$I_{xw}\dot{\omega}_{fr} = p_{f,r}T - F_{x,fr}R_e, \tag{5.7}$$
$$I_{xw}\dot{\omega}_{rl} = p_{r,l}T - F_{x,rl}R_e, \tag{5.8}$$
$$I_{xw}\dot{\omega}_{rr} = p_{r,r}T - F_{x,rr}R_e, \tag{5.9}$$

with the torque distribution coefficients $p_{i,j} \geq 0$ satisfying:

$$p_{f,l} + p_{f,r} + p_{r,l} + p_{r,r} = 1. \tag{5.10}$$

In this equation, the parameters $p_{i,j}$ are assumed to be known parameters. On this basis, the yaw rate dynamics is given by the following equation:

$$I_z\dot{r} = L_f(\sin(\delta_f)(F_{x,fl} + F_{x,fr}) + \cos(\delta_f)F_{y,f}) +$$
$$+ L_a\cos(\delta_f)(F_{x,fr} - F_{x,fl}) + L_r(\sin(\delta_r)(F_{x,rl} + F_{x,rr}) \tag{5.11}$$
$$- \cos(\delta_r)F_{y,r}) + L_a\cos(\delta_r)(F_{x,rr} - F_{x,rl}).$$

These elaborations result in a mathematical description of the AGV. The objective of the next part of this chapter is to create a set of virtual sensors which enable the estimation of $F_{x,fr}$, $F_{x,fl}$, $F_{x,rr}$, $F_{x,rl}$ and T using the measurement vector, which is available:

$$y = [r, \omega_{fl}, \omega_{fr}, \omega_{rl}, \omega_{rr}]^T \tag{5.12}$$

and also using the lateral and longitudinal accelerations a_x and a_y.

5.6 Virtual Sensor Design

The developed strategy to realise virtual sensors can start with extracting the lateral forces $F_{y,f}$ and $F_{y,r}$ from the Eqs. (5.5) and (5.11), which leads to the subsequent equations:

$$F_{y,f} = \frac{1}{\cos(\delta_f)(L_f + L_r)} \left(I_z \dot{r} + p_{f,fl} F_{x,fl} + \right.$$
$$+ p_{f,fr} F_{x,fr} + p_{f,rl} F_{x,rl} + p_{f,rr} F_{x,rr} +$$
$$+ m L_r a_y \big), \tag{5.13}$$

$$F_{y,r} = \frac{1}{\cos(\delta_f)(L_f + L_r)} \left(-I_z \dot{r} + p_{r,fl} F_{x,fl} + \right.$$
$$+ p_{r,fr} F_{x,fr} + p_{r,rl} F_{x,rl} + p_{r,rr} F_{x,rr}$$
$$+ m L_r a_y \big). \tag{5.14}$$

In these equations

$$p_{f,fl} = -\sin(\delta_f)L_f - \sin(\delta_f)L_r + L_a \cos(\delta_f), \tag{5.15}$$
$$p_{f,fr} = -\sin(\delta_f)L_f - \sin(\delta_f)L_r - L_a \cos(\delta_f), \tag{5.16}$$
$$p_{f,rl} = -2\sin(\delta_r)L_r + L_a \cos(\delta_r), \tag{5.17}$$
$$p_{f,rr} = -2\sin(\delta_r)L_r - L_a \cos(\delta_r), \tag{5.18}$$
$$p_{r,fl} = L_a \cos(\delta_f), \tag{5.19}$$
$$p_{r,fr} = L_a \cos(\delta_f) \tag{5.20}$$
$$p_{r,rl} = \sin(\delta_r)L_f - \sin(\delta_r)L_r + L_a \cos(\delta_r), \tag{5.21}$$
$$p_{r,rr} = \sin(\delta_r)L_f - \sin(\delta_r)L_r - L_a \cos(\delta_r). \tag{5.22}$$

Considering the fact that $F_x = ma_x$ and then substituting (5.13) and (5.14) into (5.4) yield:

$$p_r \dot{r} = F_{x,fl} p_{x,fl} + F_{x,fr} p_{x,fr} +$$
$$+ F_{x,rl} p_{x,rl} + F_{x,rr} p_{x,rr} + a_x p_x + a_y p_y \tag{5.23}$$

where

$$p_r = I_z \sin(\delta_f - \delta_r), \tag{5.24}$$

$$p_{x,fl} = -\frac{1}{2} L_a \sin(-\delta_r + 2\delta_f) + \frac{1}{2} \sin(\delta_r) L_a +$$
$$+ \cos(\delta_r)(L_f + L_r), \tag{5.25}$$

$$p_{x,fr} = \frac{1}{2} L_a \sin(-\delta_r + 2\delta_f) - \frac{1}{2} \sin(\delta_r) L_a +$$
$$+ \cos(\delta_r)(L_f + L_r), \tag{5.26}$$

$$p_{x,rl} = -\frac{1}{2} L_a \sin(\delta_f - 2\delta_r) - \frac{1}{2} \sin(\delta_f) L_a +$$
$$L_r \cos(\delta_f - 2\delta_r) + \cos(\delta_f) L_f, \tag{5.27}$$

$$p_{x,rr} = \frac{1}{2} L_a \sin(\delta_f - 2\delta_r) + \frac{1}{2} \sin(\delta_f) L_a +$$
$$+ L_r \cos(\delta_f - 2\delta_r) + \cos(\delta_f) L_f, \tag{5.28}$$

$$p_x = -(L_f + L_r) \cos(\delta_f) \cos(\delta_r) m, \tag{5.29}$$

$$p_y = -m(\sin(\delta_f) \cos(\delta_r) L_r + \cos(\delta_f) \sin(\delta_r) L_f). \tag{5.30}$$

The state space model of AGV can be described with the Eqs. (5.6)–(5.9) and (5.13)–(5.14)

$$G(\delta_f, \delta_r)\dot{x} = B(\delta_f, \delta_r)u + E(\delta_f, \delta_r)d. \tag{5.31}$$

In this equation

$$x = [r, \omega_{fl}, \omega_{fr}, \omega_{rl}, \omega_{rr}]^T \tag{5.32}$$

and

$$u = [a_x, a_y]^T. \tag{5.33}$$

The unknown input, which is estimated by the virtual sensor, is given by the following equation:

$$d = [F_{x,fl}, F_{x,fr}, F_{x,rl}, F_{x,rr}, T]^T. \tag{5.34}$$

The system matrices are:

$$B(\delta_f, \delta_r) = \begin{bmatrix} p_x & p_y \\ 0 & 0 \\ 0 & 0 \\ 0 & 0 \\ 0 & 0 \end{bmatrix}, \tag{5.35}$$

$$E(\delta_f, \delta_r) = \begin{bmatrix} 0 & -p_{x,fl} & -p_{x,fr} & -p_{x,rl} & -p_{x,rr} \\ -p_{f,l} & R_e & 0 & 0 & 0 \\ -p_{f,r} & 0 & R_e & 0 & 0 \\ -p_{r,l} & 0 & 0 & R_e & 0 \\ -p_{r,r} & 0 & 0 & 0 & R_e \end{bmatrix}. \tag{5.36}$$

$$G(\delta_f, \delta_r) = \operatorname{diag}(p_r, 1, 1, 1, 1). \tag{5.37}$$

Because all state variables $[r, \omega_{fl}, \omega_{fr}, \omega_{rl}, \omega_{rr}]^T$ are measured, the output equation is given by

$$y = Cx, \tag{5.38}$$

with $C = I$.

In order to allow the implementation on an on-board device, the system (5.31) is discretised using the Euler methods with the sampling time $T_s = 0.01[\text{s}]$, which leads to the following equation:

$$G_k x_{k+1} = G_k x_k + B_k u_k + E_k d_k + Wwk, \tag{5.39}$$

with

$$G_k = G(\delta_{f,k}, \delta_{r,k}), \quad B_k = T_s B(\delta_{f,k}, \delta_{r,k}), \tag{5.40}$$

$$E_k = T_s E(\delta_{f,k}, \delta_{r,k}). \tag{5.41}$$

In this equation, wk stands for an exogenous disturbance vector (which includes the discretisation error) with a known distribution matrix W, while $G(\delta_{f,k}, \delta_{r,k})$, $B(\delta_{f,k}, \delta_{r,k})$ and $E(\delta_{f,k}, \delta_{r,k})$ are obtained by substituting $\delta_{f,k}$ and $\delta_{r,k}$ into the Eqs. (5.35)–(5.37), respectively.

In order to allow a deeper consideration, it is necessary to emphasise the fact that all state variables of the Eq. (5.39) are measured. Consequently, in contrast to the approaches present in current literature (see e.g. [17] and the references therein), the attention is focused on exclusively estimating d_k. Actually, because of the fact that an estimation of the state vector is unnecessary, the proposed design procedure can be less complicated.

Finally, in order to solve the virtual sensor design problem, the following innovative adaptive estimator structure is introduced:

$$\hat{d}_{k+1} = \hat{d}_k + L(G_k x_{k+1} - G_k x_k - B_k u_k - E_k \hat{d}_k). \tag{5.42}$$

In this equation, $\hat{d}_k$ denotes an estimate of d_k and L is the gain matrix of the estimator. Substituting (5.39) into (5.42) results in the subsequent equation:

$$\hat{d}_{k+1} = \hat{d}_k + L(E_k e_{d,k} - Wwk). \tag{5.43}$$

In this equation, $e_{d,k} = d_k - \hat{d}_k$ is an unknown input estimation error. The respective dynamics are governed by the equation:

$$
\begin{aligned}
e_{d,k+1} &= d_{k+1} - d_k + d_k - \hat{d}_k - LE_k e_{d,k} - LWwk \\
&= (I - LE_k) e_{d,k} + [I \ -LW] \bar{w}_k.
\end{aligned}
\tag{5.44}
$$

In this equation, $\varepsilon_k = d_{k+1} - d_k$ and $\bar{w}_k = \begin{bmatrix} \varepsilon_k \\ wk \end{bmatrix}$. Finally, the Eq. (5.44) is transformed into a compact form:

$$
e_{d,k+1} = X_k e_{d,k} + Z\bar{w}_k,
\tag{5.45}
$$

with $X_k = I - LE_k$ and $Z = [I \ -LW]$.

To make further deliberations tractable, it is assumed that $\bar{w}_k$ is bounded within an ellipsoid

$$
\bar{w}_k \in \mathcal{E}_w, \ \mathcal{E}_w = \{\bar{w} : \bar{w}^T Q_w \bar{w} \le 1\}.
\tag{5.46}
$$

This enables the formulation of the following theorem, which constitutes the main result of this section.

Theorem 5.1 *The system* (5.45) *is strictly quadratically bounded for all* E_k *and all allowable* $\bar{w}_k \in \mathcal{E}_w$ *if there exist* $N, P \succ 0$ *and* $0 < \alpha < 1$, *such the following conditions are satisfied*

$$
\begin{bmatrix}
-P + \alpha P & 0 & P - E_k^T N^T \\
0 & -\alpha Q_w & R^T \\
P - NE_k & R & -P
\end{bmatrix} \prec 0, \quad k = 0, 1, \ldots
\tag{5.47}
$$

with $R = [P \ -NW]$.

Proof Using the definitions and the fact that $\bar{w}_k^T Q\bar{w}_k \le 1$ (cf. (5.46)) it can be concluded that

$$
\bar{w}_k^T Q\bar{w}_k < e_{d,k}^T P e_{d,k} \Rightarrow e_{d,k+1}^T P e_{d,k+1} - e_{d,k}^T P e_{d,k} < 0.
\tag{5.48}
$$

where $V_k = e_{d,k}^T P e_{d,k}$ is the Lyapunov candidate function.

Consequently, using (5.45) and defining $v_k = \begin{bmatrix} e_{d,k} \\ \bar{w}_k \end{bmatrix}$ it can be shown that

$$
v_k^T \begin{bmatrix} X^T PX & X^T PX \\ Z^T PX & Z^T PX \end{bmatrix} v_k < 0.
\tag{5.49}
$$

From (5.48) it is evident that for any $\alpha > 0$

$$\alpha v_k^T \begin{bmatrix} -P & 0 \\ 0 & Q_w \end{bmatrix} v_k < 0. \tag{5.50}$$

Consequently, employing the S-procedure to the Eqs. (5.49) and (5.50) will result in the equation

$$v_k^T \begin{bmatrix} X_k^T P X_k - P + \alpha P & X_k^T P Z \\ Z^T P X_k & Z^T P Z - \alpha Q_w \end{bmatrix} v_k < 0. \tag{5.51}$$

By implementing the Schur complement this equations can lead to

$$\begin{bmatrix} -P + \alpha P & 0 & X_k^T P \\ 0 & -\alpha Q_w & Z^T P \\ P X_k & P Z & -P \end{bmatrix} \prec 0. \tag{5.52}$$

Finally, substituting

$$P X_k = P (I - L E_k) = P - P L E_k = P - N E_k \tag{5.53}$$
$$P Z = P [I - L W] = [P - P L B] = [P - N W] \tag{5.54}$$

into (5.52) will result in (5.47), which completes the proof. $\qquad\square$

Despite the incontestable appeal of the developed approach, which can be summarised by Theorem 5.1, it is impossible to employ it for obtaining a solution of (5.47), which is feasible for all $k = 0, 1, \ldots$. In order to solve the design problem, the system (5.45) can be transformed into a "Linear Parameter-Varying" (LPV) form:

$$e_{d,k+1} = \sum_{i=\{f,r\}, j=\{l,r\}} p_{x,i,j} X^{i,j} e_{d,k} + Z \bar{w}_k, \tag{5.55}$$

where

$$X^{i,j} = I - L E^{i,j} \tag{5.56}$$

with

$$E^{f,l} = T_s \begin{bmatrix} 0 & -1 & 0 & 0 & 0 \\ -p_{f,l} & R_e & 0 & 0 & 0 \\ -p_{f,r} & 0 & R_e & 0 & 0 \\ -p_{r,l} & 0 & 0 & R_e & 0 \\ -p_{r,r} & 0 & 0 & 0 & R_e \end{bmatrix}, E^{f,r} = T_s \begin{bmatrix} 0 & 0 & -1 & 0 & 0 \\ -p_{f,l} & R_e & 0 & 0 & 0 \\ -p_{f,r} & 0 & R_e & 0 & 0 \\ -p_{r,l} & 0 & 0 & R_e & 0 \\ -p_{r,r} & 0 & 0 & 0 & R_e \end{bmatrix},$$

$$E^{r,l} = T_s \begin{bmatrix} 0 & 0 & 0 & -1 & 0 \\ -p_{f,l} & R_e & 0 & 0 & 0 \\ -p_{f,r} & 0 & R_e & 0 & 0 \\ -p_{r,l} & 0 & 0 & R_e & 0 \\ -p_{r,r} & 0 & 0 & 0 & R_e \end{bmatrix}, E^{r,r} = T_s \begin{bmatrix} 0 & 0 & 0 & 0 & -1 \\ -p_{f,l} & R_e & 0 & 0 & 0 \\ -p_{f,r} & 0 & R_e & 0 & 0 \\ -p_{r,l} & 0 & 0 & R_e & 0 \\ -p_{r,r} & 0 & 0 & 0 & R_e \end{bmatrix}.$$

Consequently, the Theorem (5.1) can be reformulated as follows:

Theorem 5.2 *The system (5.55) is strictly quadratically bounded for all E_k and all allowable $\bar{w}_k \in \mathscr{E}_w$, if N, $P \succ 0$ exist and if $0 < \alpha < 1$, such that the subsequent conditions can be satisfied:*

$$\begin{bmatrix} -P + \alpha P & 0 & P - (E^{i,j})^T N^T \\ 0 & -\alpha Q_w & R^T \\ P - N E^{i,j} & R & -P \end{bmatrix} \prec 0,$$
$$i = \{f, r\}, \ j = \{l, r\}, \tag{5.57}$$

with $R = \begin{bmatrix} P & -NW \end{bmatrix}$.

As a final step, the design procedure of virtual sensors can be consolidated to:

Off-line:

1. Select Q_w in (5.46).
2. Select $0 < \alpha < 1$ and obtain the gain matrix L of (5.42) by solving (5.57) and then substituting $L = P^{-1} N$.

On-line:

1. Set $\hat{d}_0$ and $k = 0$.
2. Obtain $\hat{d}_{k+1}$ with (5.42).
3. Set $k = k + 1$ and go to *Step 1*.

5.6.1 Uncertainty Intervals

The central objective of this subsection is to expand the virtual sensor algorithm, which was proposed in the preceding section, with an uncertainty interval that quantifies the quality of the estimates which had been achieved. Consequently, the resulting uncertainty interval provides the knowledge concerning d_k in the following form:

$$\underline{d}_k \leq d_k \leq \bar{d}_k. \tag{5.58}$$

Therefore, the objective of the subsequent part of this subsection is to present a computational framework which is capable of calculating $\bar{d}_k$ and $\underline{d}_k$. In order to solve this problem, the subsequent lemma may be taken into consideration [1]:

Lemma 5.1 *If the system* (5.44) *is strictly quadratically bounded for all* $\bar{w}_k \in \mathbb{E}_w$, *then there exists* $\alpha \in (0, 1)$ *such that*

$$V_k \leq \zeta_k(\alpha), \quad k = 0, 1, \ldots, \tag{5.59}$$

where the sequence $\zeta_k(\alpha)$ *is defined*

$$\zeta_k(\alpha) = (1 - \alpha)^k V_0 + 1 - (1 - \alpha)^k, \quad k = 0, 1, \ldots. \tag{5.60}$$

Finally, considering the fact that the estimation error $e_{d,k}$ lies within an ellipsoid (5.59), the bounds of this ellipsoid are shaped by:

$$- \left(\zeta_k(\alpha) \, c_i^T P^{-1} c_i \right)^{\frac{1}{2}} \leq e_{d,i,k} \leq \left(\zeta_k(\alpha) \, c_i^T P^{-1} c_i \right)^{\frac{1}{2}}, \quad i = 1, \ldots, n_d, \tag{5.61}$$

This result leads to the final form of the uncertainty intervals:

$$\underline{d}_{i,k} = \hat{d}_{i,k} - \left(\zeta_k(\alpha) \, c_i^T P^{-1} c_i \right)^{\frac{1}{2}}, \tag{5.62}$$

$$\bar{d}_{i,k} = \hat{d}_{i,k} + \left(\zeta_k(\alpha) \, c_i^T P^{-1} c_i \right)^{\frac{1}{2}}, \quad i = 1, \ldots, n_d. \tag{5.63}$$

From (5.60) it is evident that $\zeta_k(\alpha)$ converges to one, while its convergence rate depends exclusively on α, i.e. the closer $\zeta_k(\alpha)$ is to one, the better is the convergence rate. From the Eqs. (5.62)–(5.63) it can be deduced that the steady-state length of the uncertainty interval depends on P, i.e. on its diagonal elements. Thus, a natural measure, which needs to be optimised, should be trace(P). This can lead to the following strategy:

$$(\alpha, P, L) = \arg \max_{\alpha \in (0,1), \, P > 0, \, L} \text{trace}(P), \tag{5.64}$$

considering the constraints (5.57).

In order to obtain $\hat{d}_k$ with uncertainty intervals, which are as small as possible, *Step 2* of the *Off-line* phase of the algorithm proposed in the last section should include the optimization task, which is described in the preceding part. This assignment can be attained using widely available computational packages.

5.6.2 Diagnostic Principles

As stated in the first part of this chapter, the central goal is the development of virtual sensors which provide:

$$\hat{\boldsymbol{d}}_k = [\hat{F}_{x,fl,k}, \hat{F}_{x,fr}, \hat{F}_{x,rl}, \hat{F}_{x,rr}, \hat{T}]^T. \tag{5.65}$$

On the basis of these estimates the primary residual signal can be formed:

$$z_{T,k} = T_k - \hat{T}_k, \tag{5.66}$$

which can be employed as a source of knowledge about the desired torque distribution within the respective AGV. The remaining set of residuals relates to the longitudinal forces. The general idea concerns the definition of the longitudinal slip ratio, which for all wheels is resulting from the Eq. (compare [11]):

$$\sigma_{i,j} = \frac{\omega_{i,j} R_e}{v_x} - 1. \tag{5.67}$$

It can be seen in Fig. 5.1 that all four wheels are identical and that they are consist of a metal rim which is covered with a glued thin rubber strip. It is generally permitted to assume that the AGV is operating on a level stiff surface. Consequently, without losing generality, it can be assumed that $\sigma_{i,j,k} = 0$. This assumption can lead to the following relation:

$$\omega_{i,j} = \frac{v_x}{R_e}. \tag{5.68}$$

Because the actual velocity v_x of the AGV is available, it is possible to employ Eq. (5.68) for calculating the desired $\omega_{i,j}$. In a next step, the desired $\omega_{i,j}$ along with T are used for calculating the reference longitudinal forces $F_{x,i,j,k}$ employing the Eqs. (5.6)–(5.9). Both $\omega_{i,j}$ and $F_{x,i,j,k}$ are perceived as fault-free, because they are generated exclusively using the AGV model, while the actual AGV may endure various faults of mechanical nature and also unexpected working conditions like sliding surfaces, which are also perceived as faults. As a result of these considerations, the subsequent set of residuals is formed

$$z_{F,x,i,j,k} = F_{x,i,j,k} - \hat{F}_{x,i,j,k}. \tag{5.69}$$

5.7   Experimental Verification

A driving scenario was developed in order to be able to validate the design of the virtual sensor. This driving scenario comprises a typical manoeuvre in a production logistics environment. The subsequent list describes the scenario by means of illustrating the main parameters and their range of values:

Longitudinal velocity v_x it evolves from 0.556 [m/s] to 1.39 [m/s].
Longitudinal acceleration a_x it is constant and equal to 0.0083[m/s^2].
Front steering angle δ_f it evolves from $-7/36\pi$ [rad] to $7/36\pi$ [rad].
Rear steering angle δ_r it is constant and equal to zero.

In the scenario, the vehicle is accelerating, thus the velocity is rising. At the same time, the steering angle is altered leading to a curved route. During the whole manoeuvre the respective sensors measure the global acceleration, velocity and yaw rate. The outcome of this experimental verification are shown in the next section of this chapter.

5.8 Experimental Results and Discussion

The initial considerations concern an *Off-line* phase of the developed algorithm, which involves the Eq. (5.64) under the constraints (5.57). As a result, the optimal gain matrix of the virtual sensor (5.42) is:

$$
L = \begin{bmatrix}
-2.9501 & -27.9812 & -28.3581 & -28.3751 & -28.4762 \\
-16.3261 & 940.2042 & -311.4692 & -311.2939 & -311.1626 \\
-16.3437 & -311.0201 & 939.9038 & -311.4558 & -311.1203 \\
-16.3236 & -311.0833 & -311.4491 & 939.9822 & -311.1366 \\
-16.3238 & -310.8944 & -311.2802 & -311.2160 & 939.6778
\end{bmatrix}, \quad (5.70)
$$

along with $\alpha = 0.9$.

The next part of this section describes the experimental results from the application of the proposed virtual sensors in the driving scenario of the AGV, which has been explained in the preceding section, and the application of these virtual sensors to fault diagnosis according to the principles, which have been explained in Sect. 5.6.2. Figures 5.5, 5.6, 5.7 and 5.8 compare the longitudinal forces resulting from the proven model (nominal case) with their counterparts, which are obtained with a set of measurements that are available from the AGV sensors. Similarly, Fig. 5.9 compares the torque resulting from the proven model (nominal case) with its counterpart, which is also obtained with a set of measurements that are available from the AGV sensors. Additionally, the plots contain the uncertainty intervals calculated employing the Eqs. (5.62)–(5.63).

The first observations may be the fact that the resulting estimates are consistent with the longitudinal forces obtained employing the proven model and the fact that the forces are nearly equally distributed among the wheels. Figure 5.10 contains the residual $z_{x,fl}$. It is clearly visible that this residual is close to zero for the fault-free case. This is clearly indicating the fault-free, nominal case. The same conditions are present for the three remaining wheels and also the torque, but for the sake of an easier overview, they are omitted.

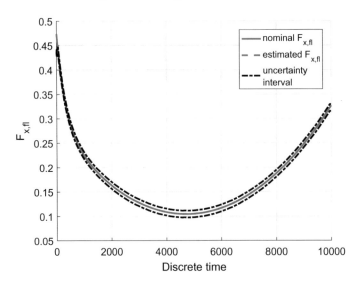

Fig. 5.5 Nominal and estimated $F_{x,fl}$

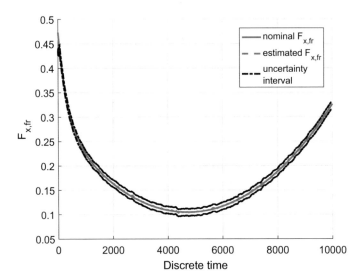

Fig. 5.6 Nominal and estimated $F_{x,fr}$

It could be shown that the virtual sensors operate properly in the fault-free case. In the next step, the performance of the virtual sensors should be assessed in a situation within which a fault is present. For that purpose, the AGV was driven towards

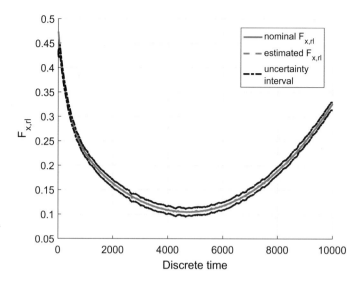

Fig. 5.7 Nominal and estimated $F_{x,rl}$

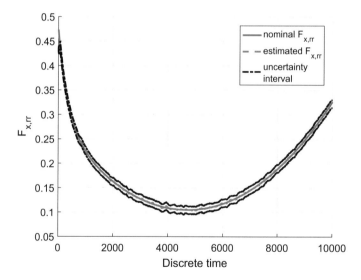

Fig. 5.8 Nominal and estimated $F_{x,rr}$

two overlapping surfaces leading to the situation that one of the wheels had no floor contact. In particular, the front right wheel was hanging in the air, therefore it did not generate the longitudinal force correctly. This unappealing phenomenon was immediately indicated by the residual $z_{x,fr}$ which is shown in Fig. 5.11.

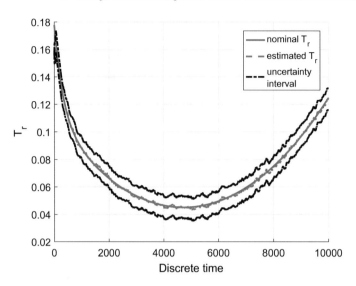

Fig. 5.9 Nominal and estimated T_r

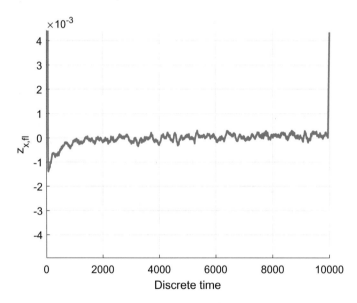

Fig. 5.10 Residual $z_{x,fl}$ for the fault-free case

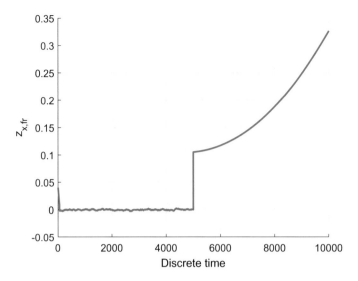

Fig. 5.11 Residual $z_{x,fr}$ for the faulty case

5.9 Conclusions

The central research question to be answered was "*how can reliable virtual sensors be designed which should measure the longitudinal forces and torques of an AGV and which are based on a quadratic boundedness approach*". For this purpose, a novel approach using quadratic boundedness was proposed, which can include bounded disturbances. This approach was implemented on a prototype AGV under development at the Hochschule Ravensburg-Weingarten for validation purposes. This AGV disposes of an unique design which allows unlimited manoeuvring possibilities and enables driving on uneven surfaces. The unique steering mechanism realises these specific qualities, but requires reliable sensor information. The developed approach to design virtual sensors does not rely on sophisticated and somewhat unreliable tire models. This fact is a considerable advantage in comparison to several approaches which can be found in literature. The use of such tire models can impair the estimation performance and can have the consequence that the whole estimation problem will become a non-linear estimation problem. The application of the approach to the AGV in a fault-free and a faulty driving scenario was successful. The application made obvious that the resulting estimates correspond to the longitudinal forces which have been calculated employing a proven reference model and made clear that the estimate immediately generates residuals in situations with faults. This immediate creation of residuals allows an effective fault detection and can be a cornerstone for fault-tolerant control.

References

1. Alessandri, A., Baglietto, M., Battistelli, G.: Design of state estimators for uncertain linear systems using quadratic boundedness. Automatica **42**(3), 497–502 (2006)
2. Aouaouda, M., Chadli, S., Shi, P., Karimi, H.: Discrete-time H_ / H-inf sensor fault detection observer design for nonlinear systems with parameter uncertainty. Int. J. Robust Nonlin. Control. **25**(3), 339–361 (2015)
3. Cai, J., Ferdowsi, H., Sarangapani, J.: Model-based fault detection, estimation, and prediction for a class of linear distributed parameter systems. Automatica **66**, 122–131 (2016)
4. Ding, S.X.: Model-based Fault Diagnosis Techniques: Design Schemes. Algorithms and Tools. Springer, Berlin (2008)
5. Foo, G., Zhang, X., Vilathgamuwa, M.: A sensor fault detection and isolation method in interior permanent-magnet synchronous motor drives based on an extended Kalman filter. IEEE Trans. Ind. Electron. **60**(8), 3485–3495 (2013)
6. Kiencke, U., Nielsen, L.: Automotive Control Systems. Springer, Berlin (2000)
7. López-Estrada, F., Ponsart, J., Astorga-Zaragoza, C., Camas-Anzueto, J., Theilliol, D.: Robust sensor fault estimation for descriptor-lpv systems with unmeasurable gain scheduling functions: Application to an anaerobic bioreactor. Int. J. Appl. Math. Comput. Sci. **25**(2), 233–244 (2015)
8. Ponsart, J.-C., Theilliol, D., Aubrun, C.: Virtual sensors design for active fault tolerant control system applied to a winding machine. Control. Eng. Pract. **18**, 1037–1044 (2010)
9. Pourbabaee, B., Meskin, N., Khorasani, K.: Sensor fault detection, isolation, and identification using multiple-model-based hybrid Kalman filter for gas turbine engines. IEEE Trans. Control. Syst. Technol. **24**(4), 1184–1200 (2016)
10. Pratama, P.S., Jeong, J.H., Jeong, S.K., Kim, H.K., Kim, H.S., Yeu, T.K., Hong, S., Kim, S.B.: Adaptive backstepping control design for trajectory tracking of automatic guided vehicles. In: AETA: Recent Advances in Electrical Engineering and Related Sciences. Lecture Notes in Electrical Engineering, vol. 371, 2016 (2015)
11. Rajamani, R., Phanomchoeng, G., Piyabongkarn, D., Lew, J.Y.: Algorithms for real-time estimation of individual wheel tire-road friction coefficients. IEEE/ASME Trans. Mechatron. **17**(6), 1183–1195 (2012)
12. Schulze, L., Wullner, A.: The approach of automated guided vehicle systems. In: Proceedings of the IEEE International Conference on Service Operations and Logistics, and Informatics (SOLI '06)
13. Stetter, R., Paczynski, A.: Intelligent steering system for electrical power trains. In: Emobility Electrical Power Train. IEEEXplore, pp. 1–6. IEEE (2010)
14. Stetter, R., Witczak, M., Pazera, M.: Virtual diagnostic sensors design for an automated guided vehicle. Appl. Sci. **8**(5) (2018)
15. Wictzak, M., Stetter, R., Buciakwoski, M., Theilliol, D.: Virtual diagnostic sensors design for an automated guided vehicle. In: Proceedings of the 10th SAFEPROCESS 2018: IFAC International Symposium on Fault Detection, Supervision and Safety for Technical Processes (2018)
16. Witczak, M.: Fault Diagnosis and Fault-Tolerant Control Strategies for Non-Linear Systems. Springer, Analytical and Soft Computing Approaches (2014)
17. Witczak, M., Buciakowski, M., Puig, V., Rotondo, D., Nejjari, F.: An LMI approach to robust fault estimation for a class of nonlinear systems. Int. J. Robust Nonlin, Control (2015)
18. Wu, S., Wu, Y., Chi, C.: Development and application analysis of AGVS in modern logistics. Revista de la Facultad de Ingeniera U.C.V., **32**(5), 380–386 (2015)
19. Zhang, H.: Software Sensors and Their Applications in Bioprocess. Springer, Berlin/Heidelberg (2009)
20. Ziemniak, P., Stania, M., Stetter, R.: Mechatronics engineering on the example of an innovative production vehicle. In: Norell Bergendahl, Grimheden, M., Leifer, L., Skogstad, P., Lindemann, U. (Eds.) Proceedings of the 17th International Conference on Engineering Design (ICED'09), vol. 1, pp. 61–72 (2009)

Part III
Fault-Tolerant Design and Control
of Automated Processes

Chapter 6
Predictive Fault-Tolerant Control of Automated Processes

This section describes the predictive fault-tolerant control of complex automated processes within a real assembly system. The assembled objects are batteries for houses. It is a well-known fact that electrical energy is one important prerequisite for the current standard of living in industrialised countries. In former times, this energy was on the one hand created by means of fossil energy carriers such as coal and oil, which dispose of the disadvantages of limited availability and emission problems. On the other hand, nuclear power was used, which is characterised by considerable operation risks and unclear solutions for the long-term disposal of nuclear waste. Therefore, considerable efforts are made in order to produce electrical energy using renewable, natural sources such as sun, water and wind. However, two of these three energy sources—sun and wind—are only available at certain times of the day and in certain situations. It is therefore necessary to store the generated electrical energy for times of demand with no available sun or wind energy. The possibilities to store large amounts of electrical energy are limited; the main feasible solution are pumped-storage power stations. A distributed energy storage system which consists of a large number of batteries in individual houses can be one solution element concerning this challenge. One main advantage is that the unavoidable losses of the electrical grid can be avoided, if solar power produced on the roof of the individual house is stored in these batteries and is also used at other times in the same house.

Currently, the high price of such battery systems is a critical factor for the large-scale implementation. Automated assembly systems can help to reduce the cost of such batteries and can thus foster the large-scale implementation. One challenge in this assembly are the risks connected with batteries with high energy densities. A reliable fault-tolerant control system is a prominent prerequisite for a safe operation of such assembly systems. It is generally advisable to describe such assembly systems which dispose of a certain unavoidable uncertainty as "Discrete Event Systems" (DES—compare Sect. 2.2).

This section presents a framework based on an interval analysis approach, which along with max-plus algebra, allows describing such uncertain discrete event systems. Based on this mathematical system description, a "Model Predictive Control-based"

© Springer Nature Switzerland AG 2020
R. Stetter, *Fault-Tolerant Design and Control of Automated Vehicles
and Processes*, Studies in Systems, Decision and Control 201,
https://doi.org/10.1007/978-3-030-12846-3_6

(MPC) fault-tolerant strategy is explained. This strategy can cope with different kinds of faults concerning processing and transportation as well as faults of the automated guided vehicles (AGVs) in the assembly system. The strategy can allow to tolerate (up to a certain degree) the influence of these faults on the overall system performance. The results presented in this chapter are based on the publications [8–10].

The next section of this chapter describes the assembly planning for the automated battery assembly system (Sect. 6.1). Section 6.2 explains the modelling of this system and Sect. 6.3 discusses the developed fault-tolerant control strategy. The application of this strategy is presented in Sect. 6.4.

6.1 Assembly Planning for Automated Battery Assembly Systems

The assembly of battery systems with high energy density is characterised by some challenges. During a certain phase of the assembly—when many cells are connected but the battery management system (BMS) is not yet functional—all possible risks need to be avoided. Therefore, this phase as well as the storage of the individual battery cells need to take place in a sophisticated room, which has to dispose of reinforced walls and should be equipped with certain vents in order to be prepared for the unlikely but possible case of an explosion of the battery cells or assembly. Naturally, the presence of human beings in this room should be limited to the absolutely necessary amount. A high degree of automation is a key prerequisite in order to minimise any potential risk for human operators. Starting from a proven design for the battery assembly, the assembly system has been planned. During the clarification of requirements, it became obvious that flexibility is one major objective for this assembly system, because the battery technology is still evolving and the demand for individual houses is amongst others dependent on political decisions such as federal subsidies and energy prices. It is therefore advantageous to apply AGVs for the internal logistics instead of less flexible solutions such as conveyor systems. Another positive aspect of using AGVs for logistics is the potential for high fault-tolerance. Usually it is possible to continue the production even if one actor (one AGV) is faulty. It is possible to achieve a high level of redundancy by providing multi-purpose AGVs which can take over many different tasks of AGVs in the case of a fault. This solution is economical and also advantageous because the multi-purpose AGV is frequently in operation and the risk of long-term idle damage is reduced. Therefore, the common risks connected with cold redundancy (compare Sect. 3.4) are reduced. Amongst others, these advantages lead to a first decision to apply AGVs for the transport within the automated assembly process. A subsequent in-depth value analysis resulted in the insight that the assembly system should be operated using four different kinds of AGVs and should consist of two nearly independent assembly cycles. It was found, the it is advantageous to operate the first cycle with two different types of AGVs (type 1 and type 2). The objective of this cycle

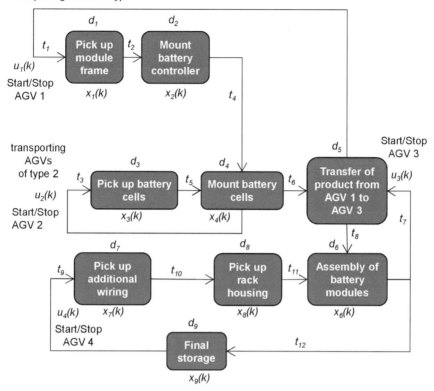

Fig. 6.1 Overview of the assembly system

is to assemble the battery frames. For the second assembly cycle also two different kinds of AGVs (type 3 and type 4) were found to be appropriate. This cycle leads to the fully assembled battery system. An overview of this system is given in Fig. 6.1.

The sequence of the first assembly cycle starts with an AGV type 1 and an AGV type 2 in their respective starting positions. From its initial position an AGV type 1 moves to the first assembly step. It moves to the frame storage and picks up an empty battery module frame. Afterwards, the safety critical battery module controller is assembled into this frame. In the next step, this AGV transports the battery frame to the so-called cell mounting system. The individual basic Lithium-ion cells are transported by an AGV type 2 from the cell storage to the cell mounting system. Through intelligent design it has been possible to make sure that this AGV can transport the number of individual cells needed to produce one battery system in only one cycle. In the next step, an appropriate number of these individual cells is assembled into the battery frame in order to achieve the desired voltage of the battery system. When this assembly is finished, the AGV type 2 can return to its starting position. The AGV type 1 can transport this assembly of the battery module already containing all battery cells and the module controller to a meeting position with the AGV type 3. Until this phase all operations have to be carried out with in the specially protected area. The final assembly can be realised outside of this room, because now the module controller, which is an integral part of the battery protection system, is already operational. In the next step, the pre-assembled battery modules are transported to the outside of the room. This has been found to be advantageous, if it is realised by transferring the pre-assembled module from an AGV type 1, which operates together with the AGVs type 2 inside the room, which is protected against the effects of explosion, to an AGV type 3, which operates together with AGVs type 4 outside of the room. The handover from AGV type 1 to AGV type 3 is made in a dedicated lock position, within which both AGVs need to have a parallel orientation. As a final step of the first cycle, also the AGV type 2 returns to its starting position. Similar to the first assembly cycle, the sequence of the second assembly cycle starts with the AGVs type 3 and type 4 in their respective starting positions. In the first step, an AGV type 3 receives the pre-assembled battery module from an AGV type 1. Then the AGV type 3 carries this module to the final assembly system. An AGV type 4 drives from its starting position to the wiring pick-up station and gathers additional wiring. In the next step, an AGV type 4 drives to the storage of the rack housings in order to get a battery module housing and bring it to the final assembly station. Next, the final assembly is carried out. After this step the AGV type 3 returns to the so-called rendezvous position (with the AGVs type 1). Finally, a AGV type 4 transports the fully assembled rack to the final storage. Then this AGV type 4 returns to its initial position. Because of an economical design, the storages which contain cell packages and frames, may only be accessed by one AGV at the same time.

6.2 Modelling the Assembly System

It is possible to describe the behaviour of this assembly system with conventional algebra using the following state and output equations:

$$
\begin{aligned}
x_1(k+1) &= max(x_1(k)+d_1, u_1(k)+t_1),\\
x_2(k+1) &= max(x_1(k+1)+d_1+t_2, S-2(k)+d_2),\\
x_3(k+1) &= max(x_3(k)+d_3, u_2(k)+t_2)\\
x_4(k+1) &= max(x_2(k+1)+d_2+t_4, x_3(k+1)+d_3+t_5, x_4(k)+d_4)\\
x_5(k+1) &= max(x_4(k+1)+d_4+t_6, x_5(k)+d_5, u_3(k)+t_2)\\
x_6(k+1) &= max(x_5(k+1)+d_5+t_8, x_6(k)+d_6, x_8(k+1)+d_8+t_{11})\\
x_7(k+1) &= max(x_7(k)+d_7, U_4(k)+t_9)\\
x_8(k+1) &= max(x_7(k+1)+d_7+t_{10}, x_8(k)+d_8)\\
x_9(k+1) &= max(x_6(k+1)+d_6+t_{12}, x_9(k)+d_9)\\
y(k) &= x_5(k)+d_5
\end{aligned}
$$

Table 6.1 The nominal and interval processing and transporting times

	Nominal time (min)	Interval time (min)
d_1	6	[5, 7]
d_2	3	[2, 4]
d_3	5	[4, 6]
d_4	8	[7, 9]
d_5	3	[2, 4]
d_6	5	[4, 6]
d_7	2	[1, 3]
d_8	4	[3, 5]
d_9	3	[2, 4]
t_1	4	[3, 5]
t_2	4	[3, 5]
t_3	2	[1, 3]
t_4	4	[3, 5]
t_5	4	[3, 5]
t_6	4	[3, 5]
t_7	4	[3, 5]
t_8	2	[1, 3]
t_9	3	[2, 4]
t_{10}	4	[3, 5]
t_{11}	3	[2, 4]
t_{12}	4	[3, 5]

However, it is rather difficult to use this state space model for the purpose of control. Additionally, it is problematic that the transportation times and the production times are not known precisely. An in-depth analysis of the assembly system resulted in the insight that it is possible to describe these times employing confidence intervals (compare Table 6.1).

Consequently, a more compact description of the system has been developed, which is able to cope with uncertainties present in the state and output equations. As a first step, the the max-plus algebra was applied, because many equations that are used to describe the behaviour of applications in the fields of control theory, machine scheduling, etc. are non-linear in conventional algebra but become linear in max-plus algebra.

6.2.1 Max-plus Algebra

The (max, +) algebraic structure $(\Re_{max}, \oplus, \otimes)$ is defined employing the following equations:

$$\mathfrak{R}_{max} \triangleq \mathfrak{R} \cup \{-\infty\},$$
$$\forall a, b \in \mathfrak{R}_{max}, a \oplus b = \max(a, b), \tag{6.1}$$
$$\forall a, b \in \mathfrak{R}_{max}, a \otimes b = a + b.$$

In these equations $\mathfrak{R}_{max}$ is the field of real numbers. The operator $\oplus$ stands for the (max, +) algebraic addition and the operator $\otimes$ for the (max, +) algebraic multiplication. The central properties of the (max, +) algebra operators can be described with the following equations:

$$\forall a \in \mathfrak{R}_{max} : a \oplus \varepsilon = a \text{ and } a \otimes \varepsilon = \varepsilon,$$
$$\forall a \in \mathfrak{R}_{max} : a \otimes e = a. \tag{6.2}$$

In these equations, $\varepsilon = -\infty$ and $e = 0$ are the neutral elements for the (max, +) algebraic addition operator and the (max, +) algebraic multiplication operator. For the matrices $X, Y \in \mathfrak{R}_{max}^{m \times n}$ and $Z \in \mathfrak{R}_{max}^{n \times p}$ the following equations are valid:

$$(X \oplus Y)_{ij} = x_{ij} \oplus y_{ij} = \max(x_{ij}, y_{ij}), \tag{6.3}$$

$$(X \otimes Z)_{ij} = \bigoplus_{k=1}^{n} x_{ik} \otimes z_{kj} = \max_{k=1,\dots,n} (x_{ik} + z_{kj}). \tag{6.4}$$

Further definitions and details concerning the (max, +) algebra formalism are contained in the publications [3, 5]. For uncertain systems the max-plus algebra can be expanded to an interval max-plus algebra.

6.2.2 Interval Max-plus Algebra

Cechlarova [6] firstly presented a methodology, which can deal with uncertainties present in the state and output equations—however this methodology was only applied for the analysis of uncertain production systems and not for control tasks [6]. This approach will be further employed for robust fault-tolerant control purposes. The (imax, +) algebraic structure ($\mathscr{I}(\mathbb{R}_{max}), \oplus, \otimes$) is defined employing the following equations:

- $\mathscr{I}(\mathbb{R}_{max})$ is a set of real compact intervals of the form $a = [\underline{a}, \overline{a}]$,
- $\forall a, b \in \mathscr{I}(\mathbb{R}_{max}), a \oplus b = \max(\overline{a}, \overline{b})$,
- $\forall a, b \in \mathscr{I}(\mathbb{R}_{max}), a \otimes b = [\underline{a} + \underline{b}, \overline{a} + \overline{b}]$.

Similar to the precedent considerations, for the matrices $X, Y \in \mathscr{I}(\mathbb{R}_{max})^{m \times n}$ and $X \in \mathscr{I}(\mathbb{R}_{max})^{n \times p}$ the following equations are valid:

$$(X \oplus Y)_{ij} = x_{ij} \oplus y_{ij} = \max(\bar{x}_{ij}, \bar{y}_{ij}) \tag{6.5}$$

$$(X \otimes Z)_{ij} = \bigoplus_{k=1}^{n} x_{ik} \otimes z_{kj} = \\ \max_{k=1,\dots,n}(\bar{x}_{ik} + \bar{z}_{kj}), \tag{6.6}$$

for all i, j.

6.2.3 Interval Max-plus Linear Model

The framework presented in the last section allows to expand the frequently applied max-plus linear state space model to interval matrices. This leads to a compact form:

$$x(k + 1) = A \otimes x(k) \oplus B \otimes u(k) \tag{6.7}$$

$$y(k) = C \otimes x(k) \tag{6.8}$$

where

- $x(k) \in \mathbb{R}_{max}^n$ is the n-dimensional state vector of $x_i(k)$, $i = 1, \dots, n$;
- $y(k) \in \mathscr{I}(\mathbb{R}_{max}^r)$ is the m-dimensional output vector of $y_i(k)$, $i = 1, \dots, m$;
- $u(k) \in \mathbb{R}_{max}^r$ is the r-dimensional input vector of $u_i(k)$, $i = 1, \dots, r$;
- $A \in \mathscr{I}(\mathbb{R}_{max}^{n \times r})$ is the state transition matrix;
- $B \in \mathscr{I}(\mathbb{R}_{max}^{n \times n})$ is the control matrix;
- $C \in \mathscr{I}(\mathbb{R}_{max}^{m \times n})$ is the output matrix.

The matrices, which can be obtained by substituting processing and transportation time intervals (compare Table 6.1) into their symbolic counterparts, are:

$$A = \begin{bmatrix}
[5,7] & \varepsilon & \varepsilon & \varepsilon & \varepsilon & \varepsilon & \varepsilon & \varepsilon & \varepsilon \\
[13,19] & [2,4] & \varepsilon & \varepsilon & \varepsilon & \varepsilon & \varepsilon & \varepsilon & \varepsilon \\
\varepsilon & \varepsilon & [4,6] & \varepsilon & \varepsilon & \varepsilon & \varepsilon & \varepsilon & \varepsilon \\
[18,28] & [7,13] & [11,17] & [7,9] & \varepsilon & \varepsilon & \varepsilon & \varepsilon & \varepsilon \\
[28,42] & [17,27] & [21,31] & [17,23] & [2,4] & \varepsilon & \varepsilon & \varepsilon & \varepsilon \\
[31,49] & [20,34] & [24,38] & [20,30] & [5,11] & [4,6] & \varepsilon & \varepsilon & \varepsilon \\
\varepsilon & \varepsilon & \varepsilon & \varepsilon & \varepsilon & \varepsilon & [1,3] & \varepsilon & \varepsilon \\
\varepsilon & \varepsilon & \varepsilon & \varepsilon & \varepsilon & \varepsilon & [5,11] & [3,5] & \varepsilon \\
[38,60] & [27,45] & [31,49] & [27,41] & [12,22] & [11,17] & [17,31] & [15,25] & [2,4]
\end{bmatrix},$$

$$B = \begin{bmatrix}
[3,5] & \varepsilon & \varepsilon & \varepsilon \\
[11,17] & \varepsilon & \varepsilon & \varepsilon \\
\varepsilon & [1,3] & \varepsilon & \varepsilon \\
[16,26] & [8,14] & \varepsilon & \varepsilon \\
[26,40] & [18,28] & [3,5] & \varepsilon \\
[29,47] & [21,35] & [6,12] & [9,19] \\
\varepsilon & \varepsilon & \varepsilon & [2,4] \\
\varepsilon & \varepsilon & \varepsilon & [6,12] \\
[36,58] & [28,46] & [13,23] & [16,30]
\end{bmatrix},$$

$$C = [\varepsilon, \varepsilon, \varepsilon, \varepsilon, \varepsilon, \varepsilon, \varepsilon, \varepsilon, [3,4]].$$

$$\tag{6.9}$$

In order to be able to describe the full functionality, it is necessary to provide a set of constraints. The system constraints can be formulated in the following form:

- Firstly, the designed system needs to follow some predefined trajectory which may be defined in the form of the follwing scheduling constraints:

$$x_j(k) \leq t_{ref,j}(k), \quad j = 1, \ldots, n. \tag{6.10}$$

In this constraint, $t_{ref,j}(k)$ is the upper bound of $x_j(k)$ at time k.
- The second constraint concerns the performance of the respective AGV,

$$\underline{u}_i \leq u_i(k) \leq \bar{u}_i, \quad i = 1, \ldots, r. \tag{6.11}$$

- The last constraint concerns the change rate:

$$u_j(k+1) - u_j(k) \geq z_j, \quad j = 1, \ldots, r. \tag{6.12}$$

In this constraint, $z_j > 0$ is the upper bound of the change rate.

On the basis of these constraints and a system description with the imax-plus algebra, a control strategy can be developed, which allows an optimal performance of the automated system.

6.3   Fault-Tolerant Control Strategy

Faults in this kind of automated process can be caused by the AGVs (AGV faults) and by elements of the assembly process (process faults). AGV faults can result from energy problems, mechanical issues (tire problems, problems of drive motors, etc.) and infrastructure issues (inappropriate surfaces, charging problems, problems with new obstacles, etc.). Process faults are usually consequences of problems in the assembly processes which are caused by the high complexity of the battery system to be assembled. This complexity implies that imperfections of the manipulators, the used materials and the preprocessing of threats or fits cannot completely be avoided. The process faults become apparent, when the production schedule is significantly violated, i.e. when production times are not achieved. For current industrial implementations constraints and control quality measures are necessary because of the common complexity of the products and processes. In general, "Model Predictive Control" (MPC) [4] is a promising candidate to address these problems, because MPC is able to handle constraints. The proposed framework can be understood as an extension of the general idea of the so-called MPC for max-plus linear systems [7].

The main element of the extension is the introduction of the interval max-plus (imax-plus) framework. This framework can be used to compute control strategies for uncertain systems and processes and for enabling appropriate fault-tolerant mechanisms. Based on the system description and fault scenarios, a suitable cost function can be derived. It is necessary to find the input sequence $u(k), \ldots, u(k + N_p - 1)$ that minimises the cost function $J(u)$:

$$J(u) = - \sum_{j=0}^{N_p-1} \sum_{i=1}^{r} q_i u_i (k + j). \tag{6.13}$$

In this equation, $q_i > 0, i = 1, \ldots, r$ is a positive weighting constant corresponding to the relative importance of the energy consumption of ith AGV, while N_p stands for the prediction horizon.

Any violation of a scheduling constraint means a faulty behaviour of the assembly system. Consequently, the optimization problem under constraints is impossible to solve in the case that the scheduling constraints are violated. In this case, it is sensible to relax the scheduling constraints as follows:

$$x_j(k) \leq t_{ref,j}(k) + \alpha_j, \quad j = 1, \ldots, n, \tag{6.14}$$

where $\alpha_j \geq 0, j = 1, \ldots, n$ should be possibly small in order to exhibit a small divergence from the intended time schedule. For the purpose of obtaining the optimal values of α_j, an altered cost function is proposed:

$$J(\alpha) = \sum_{i=1}^{n} \alpha_i, \tag{6.15}$$

and therefore a new optimization framework can be formulated:

$$J(u, \alpha) = (1 - \beta) J(u) + \beta J(\alpha). \tag{6.16}$$

In this formulation, $1 \leq \beta \leq 0$ denotes a constant which is set by the designer and which can be adjusted in order to reflect a larger importance of either $J(u)$ or $J(\alpha)$, respectively.

Given an initial condition $x(k)$, it is possible to achieve the optimal input sequence $\tilde{u}(k)^*$ by means of solving:

$$\tilde{u}(k)^* = \arg \min_{\tilde{u}(k), \alpha} J(u, \alpha), \tag{6.17}$$

for the system with faults under the constraints (6.14), (6.11) and (6.12).

It is important to note that a FTC system has to be integrated within the existing planning and control systems of an industrial company. These systems are inevitable for achieving satisfactory effectiveness and efficiency of the production processes. Usually, on the highest level an "Enterprise Resource Planning" (ERP) system is implemented, which allocates certain aspects such as raw and intermediate material and production capacity. In general, the notion ERP refers to the entrepreneurial task of planning and managing resources such as capital, human resources, operating resources, materials as well as information and communication technology according to the stated purpose of the enterprise in a timely manner (compare [1]). This system is frequently accompanied by a "Manufacturing Execution System" (MES), which is a process-related level of a multi-layered production management system and manages, for instance, the production scheduling. A typical MES provides the subsequent functionalities (compare [1, 2]):

- definition of a detailed production schedule for each product,
- definition of a detailed production process for each product,
- planning of resources (equipment, materials, human resources) for the production process for each product,
- allocating resources to the current process,
- enabling performance analysis and quality management and
- enabling information management and local production controls centres on the shop-floor.

The layers of a multi-layered production management system are illustrated in Fig. 6.2.

For the purpose of enabling fault-tolerant control, the MES can provide the current state $x(k)$ and the scheduling constraints t_{ref} for the entire prediction horizon N_p to the FTC system. The MES can also transmit the actual time at which AGVs reach individual assembly systems $u_f(k)$ as well as intervals related to the current measurements of processing and transportation times. The fault diagnosis block has the main objective to detect abnormal behaviour of the system. In this case the matrices A, B, C are replaced by their faulty counterparts A_f, B_f and C_f by the Model Alternation Module (Fig. 6.3).

The FTC system together with the Model Alternation Module perform the developed approach which can be described in detail by the FTC Algorithm (compare [8]):

Algorithm 1: FTC Algorithm

Step 0:   Initialization: Set k = 0.
Step 1:   Measurement: Measure the state $x(k)$ and the actual production and transportation $\mathbb{P} = t_1, \ldots, t_{n_t}, d_1, \ldots, d_{n_d}$.
Step 2:   Solve the linear programming problem (6.17) under the constraints (6.14), (6.11) and (6.12).

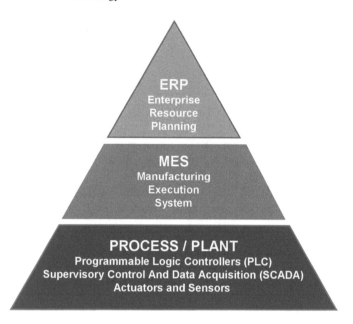

Fig. 6.2 Multi-layered production management system

Fig. 6.3 Predictive FTC
scheme

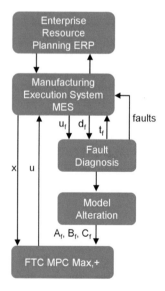

Step 3:   Use the first vector element of $\tilde{u}(k)^*$ (i.e. $u(k)^*$) and feed it into the system.
Step 4:   AGV fault diagnosis: If $s_i > \delta_i$, then the ith AGV is faulty where the residual is:

$$s_i = u_f(i, k) - u(i, k)^* \qquad (6.18)$$

for all $i = 1, \ldots, r$ and $\delta_i > 0$ being a small positive constant that is AGV-dependent and should be set by the designer.
Step 5:   Production fault diagnosis. On the basis of a set of measurements and a symbolic form of (6.7) and (6.8), calculate A_f, B_f and C_f; if $A_f \subsetneq A$ or $B_f \subsetneq B$ or $C_f \subsetneq C$ then there is a production fault
Step 6:   Model alternation: In the case of a production fault replace A, B, C with their faulty counterparts A_f, B_f and C_f; else if the ith AGV is faulty then replace B by B_f with:

$$b_{f,j,i} = b_{j,i} \otimes s_i, \quad j = 1, \ldots, m. \qquad (6.19)$$

Step 7:   Set $k = k + 1$ and go to *Step 1*.

For the validation of this innovative FTC strategy, the battery assembly system described in Sect. 6.1 was used. For this purpose, a general design procedure was developed (compare [8]):

1. Determine the system input, states and outputs.
2. Design the schematic structure of the system (compare Fig. 6.1).
3. Determine all characteristic operation times of the system, i.e., processing and transportation times (compare Table 6.1).
4. Determine the interval max plus model of the system (6.7) and (6.8).
5. Define scheduling, actuator performance (e.g. AGVs) and rate change constraints (6.14), (6.11) and (6.12).
6. Define a set of faults that may possibly affect the system, i.e. process and actuator faults (e.g. faults of AGVs).
7. Set the parameters of the FTC algorithm, i.e., N_p, q_i, δ_i, and β.
8. Run the FTC algorithm.

6.4 Application of the Fault-Tolerant Control Strategy

This section discusses the validation of the reliability of the proposed algorithm applied to the battery assembly system. During the analysis of the assembly two prominent fault scenarios were identified:

- *Case 1*: an unpermitted delay of the processing time for the assembly of the battery controller d_2, which starts at the $k = 3$ event counter.
- *Case 2*: the fault of case 1 and simultaneously a second fault of an AGV which leads to a delay by 4, which starts at the $k = 5$ event counter.

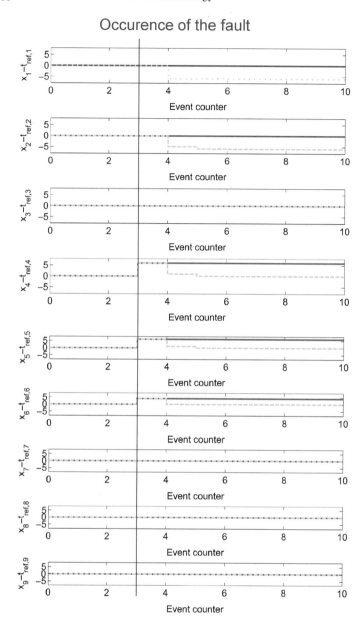

Fig. 6.4 Difference between the actual state $x_i(k)$ and the reference trajectory with FTC (green dashed line) and without it (red solid line) for *Case 1*

Fig. 6.5 Final bounds of the product outlet *y* with FTC (green dashed line) and without it (red solid line) for *Case 1*

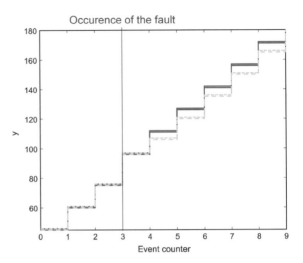

Detailed simulations were carried out on the basis of consciously defined parameters and constraints [8]. Figure 6.4 shows the result of the simulation for *Case 1*.

Figure 6.4 shows the difference between all actual components of the state, i.e. the time instant at which the different processing units start and the reference schedule. Both strategies (with and without FTC) perform equally well until the occurrence of the fault at the $k = 3$ event counter. However, after the occurrence of a fault, the MPC without FTC mechanism leads to a permanent delay behind the desired production schedule. This is obvious in the states x_4, x_5 and x_6. It is clearly visible that the proposed FTC algorithm achieves that the actual states x_4, x_5 and x_6 tend to the desired schedule. This behaviour of the system is achieved by decreasing the initial production times. Figure 6.5 shows the final product outlet *y* for *Case 1*; this clearly exhibits the performance of the proposed approach.

The full performance of the presented approach becomes visible, if the more complicated *Case 2* is considered. The result of the simulation for *Case 2* is illustrated in Fig. 6.6.

In Fig. 6.6 the difference between all actual components of the state is again shown. Until the occurrence of the fault, both strategies exhibit the same performance. Later, after the occurrence of the fault, the application of conventional MPC without the FTC mechanism will result in a permanent delay which is identifiable in the states x_3, x_4, x_5 and x_6. The result of the application of the presented FTC mechanism

Fig. 6.6 Difference between the actual state $x_i(k)$ and the reference trajectory with FTC (green dashed line) and without it (red solid line) for *Case 2*

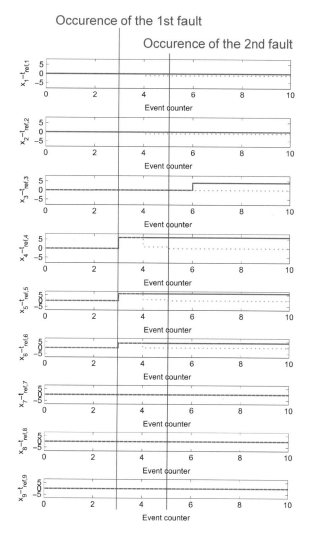

is that the states x_3, x_4, x_5 and x_6 approach the intended schedule. This is accomplished trough reduced production times x_1, x_2 and a changed arrival time for the second AGV. Figure 6.7 shows the final product outlet y for *Case 2*; this clearly exhibits the performance of the proposed approach also for more complicated cases.

Fig. 6.7 Final bounds of the product outlet y with FTC (green dashed line) and without it (red solid line) for *Case 2*

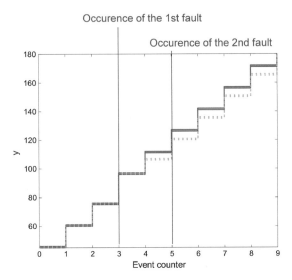

6.5 Conclusions

This chapter summarises the development of a fault-tolerant control of automated processes in an industrial battery assembly system. In the last years, the markets for renewable energy have greatly expanded, thus leading to an increased demand also for energy storage systems such as batteries. These demands can only by economically be met, if highly automated assembly systems are created. For high availability of assembly systems it is very important that such systems dispose of a high level of fault-tolerance so that they allow continuous production without disruptions. Additionally, because of the unique characteristics of batteries (e.g. the high energy density), safety issues also require an extreme level of fault-tolerance. This chapter presents an approach which can enhance the level of fault-tolerance considerably. This approach uses interval max-plus algebra, which exhibits the advantageous characteristic that it allows to describe uncertain discrete event systems. All kinds of faults of the assembly system (AGV faults and processing faults) can by addressed by the approach that was developed. The performance of the approach could be validated. Future research will aim at the FTC of assembly systems with less sensors and communication, i.e. systems with unmeasurable and therefore unavailable times which govern the system.

References

1. ISO 19440–2007 (Enterprise integration—constructs for enterprise modelling)
2. VDI 5600–2016 (Manufacturing execution systems (MES))

3. Baccelli, F., Cohen, G., Olsder, G.J., Quadrat, J.P.: Synchronization and linearity: an algebra for discrete event systems. J. Oper. Res. Soc. **45**, 118–119 (1994)
4. Blanke, M., Kinnaert, M., Lunze, J., Staroswiecki, M.: Diagnosis and Fault-Tolerant Control. Springer, New York (2016)
5. Butkovic, P.: Max-linear Systems: Theory and Algorithms. Springer (2010)
6. Cechlarova, K.: Eigenvectors of interval matrices over max-plus algebra. Discr. Appl. Math. **150**, 2–15 (2005)
7. de Schutter, T., van den Boom, T.: Model predictive control for max-plus-linear discrete event systems. Automatica **37**(7), 1049–1056 (2001)
8. Majdzik, P., Akielaszek-Witczak, A., Seybold, L., Stetter, R., Mrugalska, B.: A fault-tolerant approach to the control of a battery assembly system. Control Eng. Practice **55**, 139–148 (2016)
9. Majdzik, P., Stetter, R.: A receding-horizon approach to state estimation of the battery assembly system. In: Mitkowski, W., Kacprzyk, J., Oprzedkiewicz, K., Skruch, P. (Eds.) Trends in Advanced Intelligent Control Optimization and Automation, pp. 281–290. Springer (2017)
10. Seybold, L., Witczak, M., Majdzik, P., Stetter, R.: Towards robust predictive fault-tolerant control for a battery assembly system. Int. J. Appl. Math. Comput. Sci. **25**(4), 849–862 (2015)

Chapter 7
Prediction of the Remaining Useful Life for Components of Automated Processes

This chapter explains a strategy to employ a prognosis of the "Remaining Useful Life" (RUL) for the scheduling of automated processes in assembly systems. For many reasons, such as flexibility and availability, such assembly systems frequently employ "Automated Guided Vehicles" (AGVs) for the transportation tasks. This innovative strategy allows a consideration of the remaining operation times of the AGVs. These remaining operation times can be highly dependent on the state of the AGV supply battery, e.g. its state of charge and its health. In this chapter an innovative algorithm for the estimation of the state of the battery is explained. Additionally, a novel predictive control strategy for cooperative AGVs is presented. This strategy can allocate two alternative tasks for cooperating AGVs depending on the state of accomplished tasks from earlier steps of the assembly process. The combination of the algorithm and the strategy enables an optimum control of the assembly and transportation process in the presence of several constraints and requirements such as productivity of the individual assembly station and the whole installation or operation capabilities of the transportation AGVs. The contents of this chapter are based on the publications [37, 60].

The chapter starts with an introduction concerning the underlying research (Sect. 7.1). Section 7.2 explains the sample application and Sect. 7.3 summarises the state of the art in the three research areas *estimation of the remaining useful life, modelling the state of rechargeable batteries* and *predictive control of complex production systems with AGVs*. Approaches intended to allow a prognosis of the RUL are explained in Sect. 7.4. The algorithm applied in order to estimate the state of rechargeable batteries is elucidated in Sect. 7.5. Based on this algorithm an estimation of the remaining useful life is feasible (Sect. 7.6). The performance of the state of charge estimation is evaluated in Sect. 7.7. The framework for the predictive control of complex production systems with AGVs is in the focus of Sect. 7.8. All scientific components are verified in Sect. 7.9. The chapter is concluded with a summary (Sect. 7.10).

© Springer Nature Switzerland AG 2020
R. Stetter, *Fault-Tolerant Design and Control of Automated Vehicles and Processes*, Studies in Systems, Decision and Control 201,
https://doi.org/10.1007/978-3-030-12846-3_7

7.1 Introduction

The prevalent energy source of AGVs in production processes are rechargeable batteries such as lithium-ion batteries. The obvious advantages of this solution, such as effectiveness and efficiency, are connected with some challenges, such as availability and reliability. Some challenges result from the necessity to recharge the battery and the fact that this requires time in order to avoid very high currents and premature battery damage. During charging times of an AGV battery, the AGV usually has to remain at one place and other AGVs need to take over its tasks, if production interruptions are to be prevented. Another challenge can result from the fact that AGVs with an aged battery may even not be able to finish their respective tasks and may get "stranded" somewhere in the assembly system. This can lead to traffic jams and may require a shut-down of the production system.

The operators of manufacturing and assembly systems consequently require trustworthy estimations of the state of the AGV batteries. Such estimations would allow an adapted control of these systems and a compensation (at least partly) of the negative consequences of degraded batteries. The most important characteristics of an AGV battery are its "State Of Charge" (SOC) and its "State Of Health" (SOH) [4, 26, 27, 38, 65, 73]. The state of charge is nearly a direct consequence of the energy taken from the AGV battery since its last charge. Today it is possible to realise monitoring systems which can capture this energy by analysing the voltages and currents and their respective duration. However, one cannot always guarantee that a battery is fully charged and abrupt energy depletion can lead to higher discharge than expected. So, already an estimation of the state of charge is not that easy. Even less trivial is the estimation of the state of health. Complex electrochemical phenomena during the operating life and the direct operation cycles of a battery will decrease the usable capacity of the battery. The exact remaining capacity can only be found, if the battery is disconnected from its usual application and is continuously discharged in a controlled manner; this is impossible or at least very disturbing and complicated within the production system. These facts underline the importance of algorithms which allow estimations of the remaining usable capacity of AGV batteries.

In order to streamline the considerations concerning the SOC and the SOH, the following definition is used in this chapter, which is suitable to describe these concepts for an AGV that carries out certain tasks:

- the state of health (SOH) of a battery is a number of cycles k_f from a given initial state of charge (SOC) down to zero.

This intuitive definition gives a direct connection between SOC and SOH. In the case of two batteries, one fresh with a good SOH and one already degraded, the second one will be able to perform a smaller number of cycles k_f. Furthermore, it is obvious that the number of cycles k_f will decrease during the life-time of a battery due to the electrochemical processes mentioned above. Consequently, the number of possible cycles k_f can be employed for expressing a prediction of the "Remaining Useful Life" (RUL) [26, 38, 65] of the battery of an AGV.

Fig. 7.1 Research steps

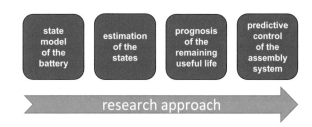

A vast amount of research concerns the control of transportation processes with AGVs and appropriate actuators and sensors [29, 61, 66]. The results are powerful control algorithms that enable very coordinated, effective and efficient transportation processes and an optimization of these processes. Up to now, the inclusion of the estimation of the state of health of the AGV battery and a prognosis of its remaining useful life has not been in the centre of these research activities.

This chapter focuses on the design of a framework which will allow a trustworthy estimation of the states of a battery and a reliable prognosis of the remaining useful life as well as a consideration of this information in the predictive planning of the transporting processes in a production system. Consequently, the research approach described in this chapter consists of the following steps:

- A state model of a battery is derived from literature.
- Algorithms are developed which allow to estimate the current states of the battery from available measurements.
- Strategies are developed which allow prognoses of the state off health and the remaining useful life based on the estimations of the state.
- A existing framework is expanded in order to allow to incorporate the information of the remaining useful life in the planning and control of the transportation processes in an assembly system.

Figure 7.1 shows these research steps.

7.2 Sample Application

The sample application are AGVs for transportation processes in a plant for the production of car seats. The main current trend in the automotive industry is autonomous driving. Especially in the crowded streets of the mega-cities of the present and future, people desire to be freed from the difficult and exhausting task of longitudinal and lateral control of their vehicle. Leading automotive companies are now accepting this challenge and plan to present more and more elaborate options of autonomous driving in the next decades. However, this will most probably change all modes of using a car and with this the whole interior of cars. The drivers do not need to concentrate on the road anymore and will be interested to spend their time for other activities such as leisure, communication and work. Drivers will desire a more relaxed position in their seat and, for communication with the passengers in the same car, they will appreciate, if the seat can turn and allow face-to-face communication. In the next decades, the traffic will be characterised by a mixture of autonomous and non-autonomous cars. Consequently, the passive safety of the drivers and passengers will still be a major issue. In the case of seats, which can turn, this leads to the necessity to integrate belts, side airbags and interaction airbags into the seat so that the belt, side airbag and interaction airbag position is still good, even if the seat is turned to a not-forward orientation. The other components of the passive safety system such as front airbags need to rely on information where, in which orientation and in which posture the driver and passengers are; sensors in the seat will contribute in the gathering of this kind of information. Another mega-trend in the car industry is electro-mobility. Due to physical limitations, the range of battery driven electrical cars will be an ongoing issue for the next decades. An astonishing large energy amount is necessary for the temperature regulation of the interior of a car. This energy can be greatly reduced, if the installations for heating and cooling are close to the human body. The largest interface to the human body is the seat, therefore heating and cooling systems in the car seats will become even more prominent in the scope of electro-mobility. These three aspects—*belt and airbag integration*, *sensor integration* and *heating/cooling integration*—will lead to heavier seats with a larger volume. This development can also cause changed requirements for the transportation systems within a seat production plant. A possibility for meeting these new requirements can be a transportation AGV with a unique steering system which has been developed at the Hochschule Ravensburg-Weingarten [58, 59, 72]. This AGV together with one of its driving modules and a view of the underside of this AGV is shown in Fig. 7.2.

The unique characteristic of this AGV is the steering system which relies on torque differences between the wheels of four independent driving modules. Theoretically, the driving modules can rotate freely with regard to a vertical axis through their respective centre (in practice, an axis brake is available which will fix a certain steering angle in the case of fast driving on a straight line in order to improve the dynamic behaviour and to increase the tolerance towards uneven sections of the floor or smaller obstacles). By balancing the torques at the two driving wheels it is possible to achieve desired steering angles for each of the modules. This unique steering system is shown in Fig. 7.3.

Kinematic and dynamic models of the AGV within the central control unit of the AGV allow the determination of desired steering angles for each module in accordance with an overall desired driving direction, orientation and driving mode of the AGV. For the estimation of the states of this AGV several measurement possibilities are present (Fig. 7.4).

The most important measurement possibilities are:

- Firstly, it is possible to measure direct product properties such as the voltage of the battery or the temperature of the drive motors.
- Secondly, product flows such as the flow of the electrical current can be measured. By means of odometry and GPS the product location and orientation can be determined.

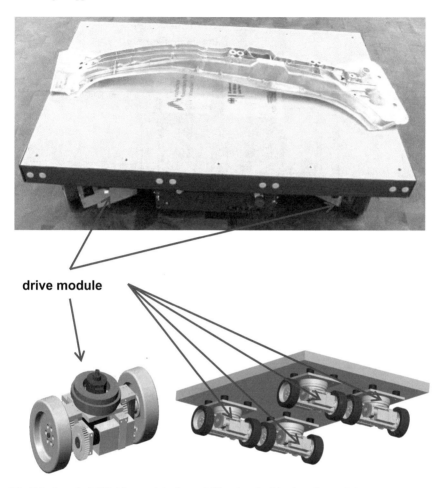

Fig. 7.2 Sample AGV, drive module (lower left) and underside view (lower right)

- Very important for the prognosis of the remaining useful life are noise, vibration and harshness (NVH) measurements.
- Other measurements can concern the product performance such as the achievable speed or the product mission fulfilment such as the number of transported goods in a certain instant of time.

Measurements concerning the battery of the AGV will be used in the later parts of this chapter. As stated above, the transportation AGVs in seat production plants need to be able to transport heavier seats with more volume. An illustrative example of the transportation of a seat frame is shown in Fig. 7.5.

The seat frame in Fig. 7.5 is not a current one, because it is currently not possible to show seat frames of future seats with safety, sensor and heating/cooling integration. The operators in seat production plants aim to use the AGVs in an effective and

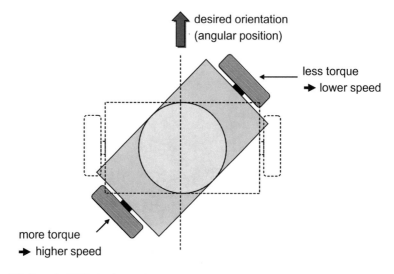

Fig. 7.3 Sample AGV: steering system

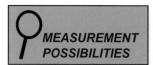

of (entity)
- product characteristics
  (e. g. voltage, motor
  temperature)
- product flows
  (e. g. electrical flow)
- product location
  and orientation
- noise, vibration, harshness
- product performance
  (e. g. speed)
- product mission fulfillment
  (e. g. transported goods)
- ...

AGV

Fig. 7.4 Sample AGV: measurement possibilities

efficient manner. For this endeavour information about the state of charge and the state of health of the AGVs battery would be very helpful. One prominent goal of the research presented in this chapter is to provide trustworthy estimates of the SOC and the SOH of the battery. These can enable predictive planning of the transportation processes and improved resource allocation. In the sample transportation scenario, an

Fig. 7.5 Sample application: transport of a seat frame

Fig. 7.6 Sample transportation scenario

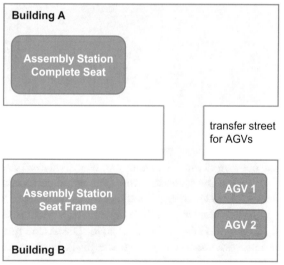

assembled seat frame is to be transported between assembly stations. This scenario is shown in Fig. 7.6.

In the station in building A (Assembly Station Seat Frame), several modules such as belt, belt lock, airbags and heating/cooling system are mounted to the metal seat frame. In another station, which is located in the especially clean building B

Fig. 7.7 Sample
transportation scenario:
process

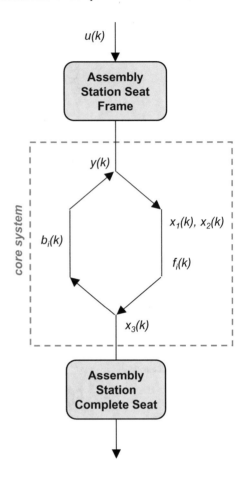

(Assembly Station Complete Seat) a specialised mechanic will add the foam parts,
seat covers and trim. A covered pathway allows AGVs to move between the buildings
A and B. In the scenario, two AGVs could perform the given transportation tasks:
AGV_1 or AGV_2. Figure 7.7 shows the process sequence.

It is important to note that only one AGV could not achieve the required schedule;
at least two AGVs need to cooperate. The assembly system can be understood as a
system with concurrency (compare [64]). In the given case, the operating range is
rather long. Additionally, the weight of the assembled seat frame can be rather high.
Therefore, the operators of the seat production plant desire a trustworthy knowledge
about the state of the battery of the AGV and assistance in the planning and controlling
of the transportation processes of the AGVs.

7.3   State of the Art

This section gives an overview of the state of the art in the three areas *estimation of prognosis of the state of health, modelling the state of rechargeable batteries* and *predictive control of complex production systems with AGVs.*

7.3.1   Prognosis of the Remaining Useful Life

Comprehensive networking, increasing digitization and new developments in the area of sensor technology enable innovative concepts and algorithms for the prognosis of the state of health of technical systems. The basis for this is the fast, efficient and comprehensive evaluation of analogue and digital data. Due to the continuous monitoring and the combination of communication, sensor technology and process data, deviations from the standard state can be detected quickly. The operators of assembly systems are given the opportunity to react early, before it comes to damage or failure. They can also plan the maintenance intervals according to the prognosis of the state of health. Condition-based maintenance can replace the conventional time-based maintenance. The optimal coordination of service operations can significantly reduce unplanned downtime, maintenance and repair costs, and increase plant uptime. Thanks to the prognosis of the state of health, future-oriented optimizations can also be made.

The core meaning of the term "prognosis" in this context is the prediction of the health of a product (compare [1, 14]). For this purpose, information concerning the prior usage of the product, the current situation and state as well as possible and probable future operation and environment conditions can be processed. For the control of complex production systems, it is very desirable to be able to rely on a trustworthy prognosis of the remaining time until a failure will force an interruption of the production process or will at least cause serious delays. Academia has reacted to this desire and a large number of research projects is addressing the general issue of health prognosis. The general goal of these research activities is to predict the "Estimated-Time-To-Failure" (ETTF) or synonymously the "Remaining Useful Life" (RUL). In a nutshell, many researchers are collecting and developing approaches which can determine, if technical systems (e.g. rechargeable batteries) show clear indicators for degradation and estimate the remaining time until the product will reach a certain failure threshold.

A failure threshold indicates that a technical system will fail to perform its intended functions. In this context, it is very important to distinguish between failure and fault. A *fault* is usually understood as a deviation of a property of the system from the acceptable conditions [23], which does not lead to permanent interruption to perform a required function [69]. This is meant by the term *failure*, which can describe catastrophic events. Usually failure thresholds can be derived from the functional requirements of a technical system (compare [19]) or from representations of product functions (compare [47]).

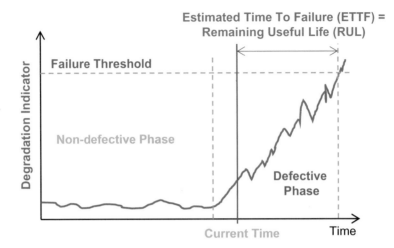

Fig. 7.8 Typical degradation process with RUL

Fig. 7.9 Bathtub curve

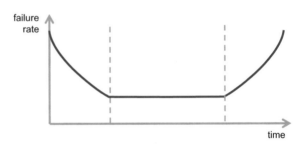

A typical ageing process of a technical system and the corresponding RUL are shown in Fig. 7.8.

It is important to note that complex systems usually consist of several modules and components and all of them can have their individual RUL. Overall, a distinct curve develops, if the failure rate is plotted over a continuous time scale (Fig. 7.9).

In this figure, three zones are clearly visible [3]. In an early zone, problems of imperfect manufacturing and assembly can cause failure (infant zone). In the main zone, a nearly constant probability of failure can be observed. When a system is aging, it will reach the late zone within which degradation effects and wear of components can lead to failure and consequently the failure rate will rise again. In addition, it should be noted that, in spite of its logical appeal, a scientific discourse can be observed, whether all zones of the bathtub curve are valid for all products and if techniques like burn-in can improve the behaviour in the early zone (compare [28]).

As stated above, several research groups worldwide are active in researching the general topic of health prediction; the results have been compared and summarised in several concise reviews [18, 24, 30, 31, 53]. On a very general level, the approaches for the direct prognosis of the RUL can be distinguished based on their central information basis: they can be either model-based, data-driven or hybrid [9]. Physical relationships such as material ageing under certain loads are the basis for model-based approaches and are formulated in mathematical representations. One important advantage of model-based approaches is the general tendency that a certain amount of extrapolation is possible when using such models and the smaller dependence on given data. Additionally, such representations allow plausibility checks for instance by means of dimension analysis. The models can incorporate knowledge concerning the life-cycle of products, about load conditions in certain life phases, about their structure and geometry and about material characteristics. This knowledge can lead to an in-depth understanding of potential physical failure mechanisms and allow the estimation of the RUL [40, 41]. A special form are knowledge based methods which employ a comparison between a system behaviour and faults, which are previously defined; a distinction between expert systems and fuzzy systems is possible [33, 55].

The starting point for data driven approaches lies in rather large amounts of condition monitoring data. These approaches rely on "Artificial Intelligence" (AI) techniques such as neuronal networks [22] and "Support Vector Machines" (SVM) [9] or statistical methods such as regressive models [42, 63] and similarity-based approaches [67]. Notably are also the trend extrapolation and "Autoregressive Moving Average" (ARMA) models [42, 43, 63] as well as the Particle Filtering [32], the "Extended Kalman Filter" (EKF) [56], the Interval Observers [70], the "Unscented Kalman Filter" (UKF) [71], Bayesian techniques [35, 51], Gaussian Process Regression [16] and Hidden Markov and Hidden Semi-Markov Models [13].

A combination of both main approaches is used for hybrid models. Figure 7.10 summarises approaches for estimating the RUL.

7.3.2   Models of Battery Degradation

Today, in many applications, lithium-ion rechargeable batteries are used, because in terms of energy density and durability they exhibit advantageous characteristics compared to lead-acid, nickel-cadmium and nickel-metal-hydride batteries. In order to make use of the advantages of the battery technology it is necessary to analyse and represent their degradation. Several scientific projects address these issues and use "Bayesian Frameworks" [50], the "DempsterShafer Theory" (DST) with the "Bayesian Monte Carlo" (BMC) method [17], "Extended Kalman Filters" (EKF) [20, 73], a dual filter consisting of an interaction of a standard Kalman filter and an "Unscented Kalman filter" (UKF) [4], "Linear Parameter-Varying" (LPV) models [48] and nonlinear predictive filters [21]. Additionally, methods of artificial intelligence are used such as "Support Vector Machines" (SVM) and "Relevance Vector Machines" (RVM) [27, 68]. With a focus on car industry, elaborate reviews of battery degradation estimations for automotive applications have been elaborated by Barre et al. [6], Rezvanizaniani et al. [49] and Berecibar et al. [7]. It is very difficult to estimate the RUL of rechargeable battery systems, because two degradation mechanisms occur simultaneously: calendar degradation and operation degradation.

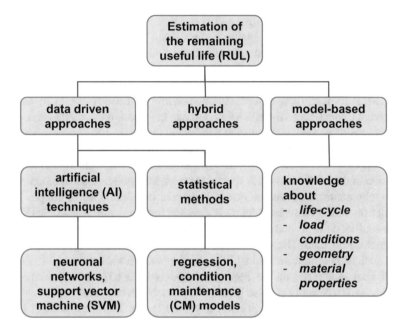

Fig. 7.10 Overview of RUL estimation approaches

The core meaning of calendar degradation is that the usable capacity of a rechargeable battery will decrease over the life-span. This degradation process is usually not linear and depends on external factors such as temperature. Operation degradation means ageing processes because of charge-discharge-cycles. The characteristics of these cycles can strongly influence the ageing; for instance, fast discharge cycles will lead to stronger degradation than slow discharge cycles. The same is true for the charging of the batteries. The result of calendar and operation degradation are an increased resistance of the battery and a decreased capacity. Five main state parameters for rechargeable battery cells could be identified by Saha et al. [50]:

- the aging parameter λ_{R_E} for the electrolyte resistance R_E
- the aging parameter $\lambda_{R_{CT}}$ for the charge transfer resistance R_{CT}
- the electrolyte resistance R_E
- the charge transfer resistance R_{CT}
- the capacity at rated current C_I

Saha et al. [50] elaborated degradation models applying exponential growth models using the following equation:

$$mpv_{IBP} = C_{IBP} \cdot exp(\lambda_{IBP} \cdot t). \tag{7.1}$$

In this equation, mpv_{IBP} denotes the model predicted value of an internal battery parameter IBP such as R_E or R_{CT}. The system model can be formulated using the following equations:

$$
x_k = \begin{cases} \lambda_{RE} : x_{1,k} = x_{1,k-1} + \omega_{1,k} \\ \lambda_{RCT} : x_{2,k} = x_{2,k-1} + \omega_{2,k} \\ R_E : x_{3,k} = x_{3,k-1} exp(x_{1,k} \cdot \Delta t) + \omega_{3,k} \\ R_{CT} : x_{4,k} = x_{3,k-1} exp(x_{2,k} \cdot \Delta t) + \omega_{4,k} \\ C_I : x_{5,k} = \alpha(x_{3,k} + x_{4,k}) + \beta + \omega_{5,k} \end{cases} \quad y_k = \begin{cases} R_E^* : y_{1,k} = x_{3,k} + v_{1,k} \\ R_{CT}^* : y_{2,k} = x_{4,k} + v_{2,k} \end{cases}
$$

$$(7.2)$$

In these equations, x_k denotes the state vector and y_k denotes the measurement vector. It is possible to compute the state of charge (SOC) using the current capacity estimate $x_{5,k}$. This approach shows good accuracy. However, the impedance measurement requires expensive and bulky equipment and is time consuming [17]. A pragmatic mathematical model is proposed by Taborelli and Onori [62]. They use the following state space formulation for a unit of four cells named: "average cell":

$$
\begin{cases} SOC_{k+1} = SOC_k - \frac{\Delta t}{Q_{nom}} I(k) \\ V_{CT}(k+1) = e^{-\frac{\Delta t}{\tau_{CT}}} V_{CT}(k) + R_{CT}(1 - e^{-\frac{\Delta t}{\tau_{CT}}})I(k) \\ V_{Dif}(k+1) = e^{-\frac{\Delta t}{\tau_{Dif}}} V_{Dif}(k) + R_{Dif}(1 - e^{-\frac{\Delta t}{\tau_{Dif}}})I(k) \end{cases}
$$

$$(7.3)$$

The dynamic response of battery cells is modelled using two resistance/capacity branches (R_{CT}, C_{CT} and R_{Dif}, C_{Dif}) and the time constants $\tau_{CT} = R_{CT}C_{CT}$ and $\tau_{Dif} = R_{Dif}C_{Dif}$, which represent "Charge Transfer" (CT) and diffusion (Dif) processes. The average output voltage $V(k)$ is related to the relevant voltage drops:

$$
V(k) = V_{OCV}(SOC(k)) - V_{CT}(k) - V_{Dif}(k) - R_0 I(k) \tag{7.4}
$$

with the average cell "Open Circuit Voltage" (OCV) function $V_{OCV}(SOC)$ (i.e. the difference of electrical potential between the two terminals of the cell when disconnected from any circuit) and the internal resistance of the battery cell R_0. Taborelli and Onori [62] define the state vector as $x(k) = [SOC(k) \quad V_{CT}(k) \quad V_{Dif}(k)]^T$. The input of the model is the current $u_l = I(l)$ and its output is the voltage $y_l = V(l)$. Resulting from this is the non-linear state space model of the battery:

$$
x_{l+1} = Ax_l + Bu_k \tag{7.5}
$$

$$
y_l = g(x_l, u_l) \tag{7.6}
$$

The respective matrices A and B are:

$$
A = \begin{bmatrix} 1 & 0 & 0 \\ 0 & e^{-\frac{\Delta t}{\tau_{CT}}} & 0 \\ 0 & 0 & e^{-\frac{\Delta t}{\tau_{Dif}}} \end{bmatrix};
$$

$$
B = \begin{bmatrix} -\frac{\Delta t}{Q_{nom}} \\ R_{CT}(1 - e^{-\frac{\Delta t}{\tau_{CT}}}) \\ R_{Dif}(1 - e^{-\frac{\Delta t}{\tau_{Dif}}}) \end{bmatrix}.
$$

$$(7.7)$$

Section 7.5 presents an innovative algorithm to estimate the state of a rechargeable battery.

7.3.3 Predictive Control of Assembly Systems

The application example in this chapter is a production system using AGVs for the transportation of seat frames and complete seats over more than one building. The main requirements for this kind of system are a high level of reliability and availability as well as a high level of flexibility, which may be required by future seat variations. Future process control strategies need to address these requirements and need to tolerate a certain level of faults in the production system. These requirements can be addressed with an active "Fault-Tolerant Control" (FTC) framework [34, 52], which can be based on a production system description as "Discrete Event System" (DES) [39, 44] and which may apply the discrete event interval max-plus algebra [5, 8]. In the last years, a model predictive control-based fault-tolerant strategy was created by applying this mathematical description (compare also Chap. 6).

7.4 Approaches for the Prognosis of the Remaining Useful Life

This sections discusses possibilities for generating a reliable prognosis of the "Remaining Useful Life" (RUL) of a technical system. Initially the elaborations concentrate on degradation indicators. Then the prognosis of the RUL for superordinate systems is discussed. Finally, the relationships between RUL and fault-tolerance are elaborated.

7.4.1 Degradation Indicators

As can be seen in Fig. 7.4, usually several possible measurements are available, which could theoretically be used for the estimation of the RUL. In order to search for such measurements and to get an overview, a method called "black box analysis" or "black box diagram" [60] can be employed. In Fig. 7.11 an example of a black box diagram of an AGV is depicted.

This technique allows a concentration on the inputs and outputs of a product. These inputs and outputs can be prominent input information for RUL estimation processes. The central output of any technical system is its performance concerning the main functions of the respective system and their quantitative fulfilment. As the main function of a transporting AGV is "transport goods" the performance can

Fig. 7.11 Black box
diagram of an AGV

electrical ⟹
energy
consumption

auxiliary ⟹
substances:
lubricants, ...

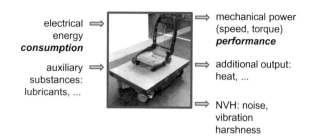

⟹ mechanical power
(speed, torque)
performance

⟹ additional output:
heat, ...

⟹ NVH: noise,
vibration
harshness

be characterised by the achievable speed and the torque at the wheels which will
determine the climbing and acceleration capabilities. The consumption of products,
such as the electrical energy consumption of AGV, can also lead to health indicators
because several forms of degradation can lead to less than perfect surfaces, to higher
friction and to increasing energy consumption. Vibrations which are caused by moving machine elements can also be a main source of health indicators, for instance
because less than perfect surfaces can also lead to increased vibration. Possible additional outputs can be, for example, heat—a higher heat emission than expected and
usual can also indicate problems within the technical system. The next subsections
will explain five prominent performance indication approaches.

7.4.1.1   Performance Oriented Degradation Detection

The performance of a product can be a prominent indicator for its ageing and health.
For an AGV the performance can be indicated by the maximum achievable speed and
the climbing capability which is connected with the torque at the wheels. In order
to monitor these ageing processes of technical systems a dimensionless degradation
indicator can be defined:

$$DI_{P_0} = \frac{current \quad perfomance}{initial \quad performance}. \tag{7.8}$$

In this equation, DI_{P_0} denotes the performance oriented degradation indicator in relation to the initial performance. The performance of many products is not degrading
linearly; possible degradation behaviours are shown in Fig. 7.12.

Usually products dispose of a better performance than the performance, which is
really needed for the given task, because certain safety margins are considered and a
certain amount of ageing can be tolerated. Typical required minimum performances
amount for between 50 and 90% of the initial performance. In the usual cases the performance degradation is not linear. It is even possible that the performance increases
after a certain time, because some manufacturing imperfections may be smoothed
by the product's operation. In nearly all cases, the performance will go down before
the event of a failure, for instance because of the erosion of friction-reducing layers.
Consequently, complex performance degradation behaviour may occur (compare
Fig. 7.12).

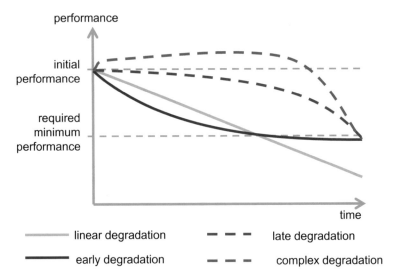

Fig. 7.12 Typical performance degradation

Performance-deteriorated parameter data and life-cycle data are also used in current research projects in order to be able to estimate the RUL [9].

7.4.1.2   Consumption Oriented Degradation Detection

This subsection deals with the detection of the degradation of a product, which is oriented to the consumption of this product or its components, e.g. the electrical power consumption of an electrical actuator. It is rather common that ageing technical systems will exhibit a decreased efficiency, usually as a consequence of less perfect surfaces within the system leading to higher friction. It is possible to define a dimensionless degradation indicator:

$$DI_{C_0} = \frac{initial\ consumption}{current\ consumption}.$$ (7.9)

In this equation, DI_{C_0} represents the consumption oriented degradation indicator. In general, one my also use the efficiency of a system as degradation indicator – this will lead to analogous results.

Today, it is usually possible to monitor the consumption of the different system elements, because this monitoring can assist the development of efficient product components and has been therefore already realised in many technical systems. Consequently, this kind of degradation detection disposes of an enormous potential for an economic detection of ageing product components.

7.4.1.3 Additional Output Oriented Degradation Detection

It can be seen in Fig. 7.11 that systems often produce several outputs in addition to the desired outputs that are necessary for the functionality of the technical system under consideration. It is possible to define a dimensionless degradation indicator:

$$DI_{AO_0} = \frac{initial\ additional\ output}{current\ additional\ output}.$$ (7.10)

In this equation, DI_{AO_0} represents the additional output oriented degradation indicator. For the electrical motors of an AGV the main additional output is heat. Usually, a temperature sensor is installed in order to protect the electrical motor from overheating; the information of this sensor can also be used in order to detect degradation. Especially, when certain additional information is given, such as the current power delivered by the electrical motor, it is possible to detect an increased heat production which could be, for instance, caused by additional friction in worn bearings of the motor, thus indicating degradation. Noise and vibration can also be understood as additional output. However, due to their enormous relevance for the prognosis of the RUL, they are discussed in a separate subsection (Sect. 7.4.1.5).

7.4.1.4 Auxiliary Substance Oriented Degradation Detection

Auxiliary substances such as oils can also be found in large-scale heavy-duty AGVs, for instance in hydrostatic bearings. Wear or damage in these bearing can result in increased losses of these auxiliary substances. It is possible to define a dimensionless degradation indicator:

$$DI_{AS_0} = \frac{initial\ loss\ of\ auxiliary\ substance}{current\ loss\ of\ auxiliary\ substance}.$$ (7.11)

In this equation, DI_{AS_0} represents the auxiliary substance oriented degradation indicator. For many systems, such as combustion engines, also the condition of auxiliary substances may be used for degradation detection, e.g. degraded lubricants. However, such analyses are only possible for very large AGVs, because otherwise no auxiliary substances are present.

7.4.1.5 Noise/Vibration/Harshness Oriented Degradation Detection

The concept of "Noise, Vibration and Harshness" (NVH), describes phenomena concerning the noise and vibration characteristics of technical systems, e.g. AGVs. A dimensionless degradation indicator can be defined as:

$$DI_{NVH_0} = \frac{initial\ NVH\ level}{current\ NVH\ level}.$$ (7.12)

In this equation, DI_{NVH_0} represents the NVH oriented degradation indicator. The NVH level can typically be measured by vibration sensors. Recent research initiatives investigate the possibility to measure NVH even without such vibration sensors. Rad et al. [46] propose an algorithm which can use the current signal of an electrical motor (in this case the spindle motor of a milling machine) and apply a time-frequency analysis in order to gather the NVH effects. This kind of analysis would also be possible in the electrical motors of an AGV; especially in the case that motor current sensors are installed for other purposes, e.g. control purposes. More general, it can be concluded that the monitoring and analysis of the input parameters of actuators, such as motors (compare [10]), may be used to detect degradation, e.g. because of increased fluctuations because of increased NVH. Additionally, already existing sensors could also be used for this purpose (this can be referred to as "added use of sensors"). For instance, an increased noise level of one sensor might also indicate an increased NVH level at a component in the proximity. Therefore, in future "actuators as sensors", "added use of sensors" and sensor fusion may be used for NVH determination; further research in this direction can be extremely fruitful.

7.4.2   Prognosis of the Remaining Useful Life of Superordinate Systems

The importance of system hierarchies has been already stated in Sect. 2.5. In general, the prognosis of the RUL of superordinate systems is similar to the determination of the reliability of complex systems. The calculation of the reliability of a technical system is based on the probability of failure λ of its individual components and the nature of their logical connection, i.e. it has to be considered, whether components are in a redundant arrangement. Obviously, the connections between these components, i.e. the interfaces of the system, also have to be considered, because they can also reduce the reliability (and also the RUL) of a system. For a prognosis of the RUL of superordinate systems, the nature of the logical connections between their subsystems and components also has to be considered, but there are additional important aspects. In theory, the RUL of a superordinate system cannot be larger than the RUL of the subsystem or component with the smallest RUL which is necessary for the main functions of the superordinate system (in this context "necessary" means that it is not a redundant element or it is an element which is only necessary for certain functions). However, in industrial practice, the RUL of the superordinate system can be greatly enhanced, if:

- subsystems and components can be maintained in order to increase their respective RUL,
- subsystems and components can be repaired in order to increase their respective RUL or
- subsystems and components can be replaced.

Table 7.1 Categories for subsystems and components

Category	Description
Category A	Subsystems and components which are crucial for the main functions of the superordinate system, which cannot be maintained, repaired or replaced with admissible technical and economic effort
Category B	Subsystems and components which can be maintained, repaired or replaced with admissible technical and economic effort, but whose maintenance, repair and replacement do require downtime of the superordinate system
Category C	Subsystems and components which can be maintained, repaired or replaced with admissible technical and economic effort and whose maintenance, repair and replacement do not require downtime of the superordinate system

This leads to the formulation of a first central insight concerning the RUL of super-ordinate systems:

When the RUL of a complex system consisting of subsystems and components is to be determined, one needs to take into account the possibility that subsystems and components can be maintained, repaired or replaced.

Additionally, a first estimation concerning the RUL of a superordinate system can be formulated:

The RUL of a superordinate system cannot be larger than the RUL of the subsystem or component with the smallest RUL which is necessary for the function of the superordinate system, if this subsystem or component cannot be maintained, repaired or replaced with admissible technical and economic effort.

These factors aggravate a determination of the RUL of superordinate systems. It is therefore sensible to reduce the number of considered subsystems and components. The reduction can be supported by the subsequent Table 7.1.

Depending on the importance of the determination, one can decide either to only take into account the subsystems and components from category A or from categories A and B. For the calculation the RUL of the superordinate system the procedure proposed for multicomponent systems as proposed by Ferri et al. [15] can be applied (Fig. 7.13).

Initially, a fault tree that represents the superordinate system with its subsystems and components needs to be generated. In a second step, this fault-tree is converted into a minimum cut sets representation. Cut sets are unique combinations of subsystem or component failures which can cause a failure of the superordinate system. A minimal cut set is achieved, if, when any basic event is removed from the set, the remaining events collectively are no longer a cut set [25]. As an simplified example, an automated process which consists of two AGVs and two "Loading/UnLoading" (LUL) systems is shown in form of these two kinds of representation in Fig. 7.14.

For a correct operation of the superordinate system—an automatic assembly process—it is required that at least one subsystem (AGV with its LUL system) is working properly and that within this system both the AGV and the LUL system have to be functional. This interaction of the system elements is shown as fault-tree (Fig. 7.14 – left side) and minimum cut sets (Fig. 7.14 – right side).

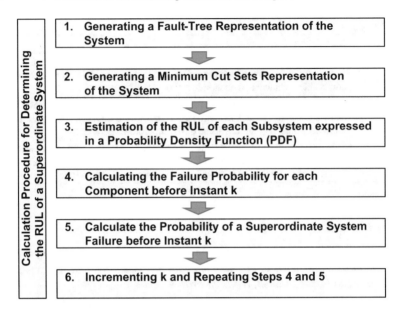

Fig. 7.13 Procedure for determining the RUL of superordinate systems

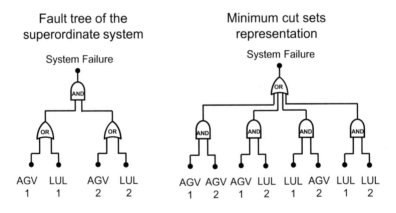

Fig. 7.14 Fault tree and minimum cut sets representation of a superordinate system

For the third step, a RUL estimation for all subsystems or components of category A or categories A and B needs to be collected (see Sect. 7.3.1) and expressed in the form of a "Probability Density Function" (PDF). It is now possible to calculate the probability of each subsystem or component to fail before instant k. Based on the minimal cut sets representation, the probability of each cut set that failure will occur before instant k can be determined employing the following equation:

$$P(c_i) = \prod_{j=1}^{n} P(e_j). \tag{7.13}$$

In this equation, $P(c_i)$ denotes the probability of the ith cut set, $P(e_j)$ stands for the probability of the e_j basic event and n denotes the number of basic events in the ith cut set.

Now, another equation can be employed in order to calculate the probability of the top event (the probability of a failure of the superordinate system) to occur before instant k:

$$P_T = 1 - \prod_{i=1}^{m} P(c_i). \tag{7.14}$$

In this equation, P_T denotes the probability of the top event and m stands for the number of cut sets. This probability of the top event expresses the probability that failure occurs in at least one cut set; this is numerically equal to one minus the probability that failure occurs in no cut set. Subsequently, steps 4 and 5 are repeated for subsequent instants. The proposed procedure results in a "Cumulative Distribution Function" (CDF) that represents the probability of a failure of the superordinate system to occur over time [45].

7.4.3 Relationships between Remaining Useful Life and Fault-Tolerance

The main topic of the book is fault-tolerance; the main topic of this chapter is the remaining useful life. One may ask, if a connection between those exists, which has to be considered in the development and operation of fault-tolerant systems. A closer analysis has led to the insight that different forms of relationship can be distinguished (Fig. 7.15).

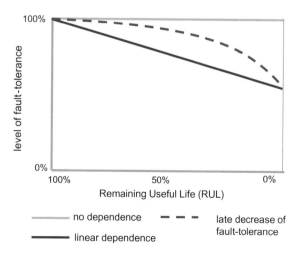

Fig. 7.15 Relationship between RUL and fault-tolerance

For certain systems and components, no dependency can be found between RUL and fault tolerance, i.e. even close to their "End-of-Life" (EoL) such systems and components still offer the same level of fault-tolerance they exhibited when they were new. This behaviour is also valid for systems and components with no fault-tolerance at all. For some kinds of systems or components, the fault-tolerance might decrease in a linear or nearly linear manner. This could be typical for AGV batteries, which can more easily cope with slight overload, if they are new, but close to the EoL this tolerance will diminish. For another kind of systems or components, the fault-tolerance will remain about the same until a phase closer to the EoL, but will decrease more intensively in this phase (late decrease of fault-tolerance – see Fig. 7.15). Such behaviour is typical of bearings, within which some hard layers have been already partly damaged and are therefore more sensible to the consequences of faults such as overload. The fact that RUL and fault-tolerance are not necessarily independent has to be considered by the design engineers and they need to assure that the level of fault-tolerance close to the EoL is still large enough to fulfil the central requirements.

This section so far has tried the answer the question: *Is there an influence from the remaining useful life of a system on the fault-tolerance of this system?* The resulting insight has been that an influence is possible and that certain cases can be distinguished. One might also ask another question: *May an increase of the fault-tolerance also lead to an increase of the RUL of this system?* Several examples can easily be identified which underline that this is also possible. Obviously redundancy is one prominent element of fault-tolerance; it is obvious that the presence of redundancy for crucial system elements can prolong the RUL of the superordinate system. Virtual sensors, which can also be part of a fault-tolerant system will usually exhibit no ageing and can consequently also prolong the RUL. Similar insights are possible when looking at the other options for increasing fault-tolerance such as "functional diversity", "physical diversity", "over-actuation" and "inherently fault-tolerant system configurations" (compare Sect. 3.4. Therefore the subsequent insight can be formulated:

Measures, which aim to increase the fault-tolerance of a system, will frequently also have a positive influence on the remaining useful life of this system.

Systems which allow active fault-tolerant control dispose of a "Fault Detection and Identification" (FDI) system. Such FDI systems rely on monitoring of certain parameters of the system—this monitoring can also be used for the early detection of required maintenance. This detection may allow a planned maintenance during scheduled down-time thus leading to a prolonged RUL of the superordinate system. This leads to another central insight, going back to approaches for the prognosis of the RUL:

Measures, which aim to increase the fault-tolerance of a system, will frequently also support approaches for an on-line prognosis of the remaining useful life of this system.

Additionally, system designers need to be aware that the fault-tolerance of systems can sometime be misused in order to prolong the operation time. For instance, a system operator may continue to operate a system within which some redundant elements have already failed. Obviously, the fault-tolerance and even safety of this system might be deteriorated. It is important that operators are informed about such events. Certain strategies, such as a emergency operation scheme with reduced system performance, may increase the interest of the operator to bring the system back to its original configuration.

This section has concentrated on approaches for the prognosis of the RUL. The next two sections concern the prognosis of the RUL of AGV batteries; the estimation of the state is an intermediate step which is discussed in the next section.

7.5 Estimation of the Battery State

The estimation of the battery state can start with the cell model described in the Eqs. (7.5), (7.6). This section aims to explain an innovative estimator employed to estimate the state of charge.

Scientific publication (for instance [54] and the cited references) report that frequently the "Extended Kalman Filter" (EKF) is applied for solving this problematic task. This selection is probably caused by the non-linearity which is present in (7.6). Despite the evident advantages of this approach, it inherits a disadvantage of the EKF, which is that it may quickly diverge in certain situations. In order to address this issue, an innovative estimator is proposed in this section and its convergence is conscientiously proven.

An ideal starting point can be the fact that the output Eq. (7.6) can be written in the following form:

$$y_1 = Cx_k + Du_k + h(x_{1,k}).$$ (7.15)

In this equation $C = [0, 1, 1]$, $D = R_0$, $h(\cot) = V_{OCV}(\cdot)$. In a next step, this description may be expanded in a manner which allows coping with measurement and process noise and/or disturbances:

$$x_{k+1} = Ax_k + Bu_k + W_1 w_k,$$ (7.16)
$$y_1 = Cx_k + Du_k + h(x_{1,k}) + W_2 v_k.$$ (7.17)

In these equations, $w_k \in \mathbb{R}^{n_w}$ and $v_k \in \mathbb{R}$ stand for the process and output noise or disturbance, respectively. Based on this, an innovative estimator structure may be proposed:

$$\hat{x}_{k+1} = A\hat{x}_k + Bu_k + K(y_k - C\hat{x}_k - Du_k - h(\hat{x}_{1,k})).$$ (7.18)

In this equation, $\hat{x}_k$ stands for the state estimate and K denotes for the estimator gain matrix. In order to deeply analyse the phenomena, the state estimation error is defined as:

$$e_{k+1} = x_{k+1} - \hat{x}_{k+1}. \tag{7.19}$$

By means of substituting (7.16), (7.18) into (7.19) the following equation can be achieved:

$$e_{k+1} = Ae_k + W_1w_k - K(Ce_k + h(x_{1,k}) - h(\hat{x}_{1,k}) + W_2v_k). \tag{7.20}$$

Here it becomes obvious that the present non-linearity causes the central challenge for proving the convergence of the estimation error:

$$h(x_{1,k}) - h(\hat{x}_{1,k}). \tag{7.21}$$

One possibility to address the issue is to carry out a linearisation; this is also done when a EKF is applied. Nevertheless, this is a crucial factor which contributes to the possible divergence of the EKF. This can be avoided by means of defining the subsequent function:

$$\gamma(x_k, \hat{x}_k) = \begin{cases} \frac{h(x_{1,k}) - h(\hat{x}_{1,k})}{x_k - \hat{x}_k} & x_k \neq \hat{x}_k \\ 0 & \text{otherwise} \end{cases} \tag{7.22}$$

For this function the Eq. (7.20) can be compressed to:

$$e_{k+1} = Ae_k + W_1w_k - K(Ce_k + \gamma(x_k, \hat{x}_k)e_k + W_2v_k). \tag{7.23}$$

For continuing this elaboration, the nature of $h_{x_{1,k}}$ (i.e., $V_{OCV}(SOC)$) has to be investigated. Two important measures can be employed to describe it:

- The state of charge (SOC) is within the range $0 < x_{1,k} < 1$;
- the function $h(\cdot)$ linking SOC and V_{OCV} is positive and monotonic.

The following bounds on $\gamma(x_k, \hat{x}_k)$ are forced by these observations:

$$0 < \gamma(x_k, \hat{x}_k) < \bar{\gamma}. \tag{7.24}$$

In these bounds, $\bar{\gamma} > 0$ denotes a constant which is dependent on the shape of $h(\cdot)$. This constant allows rewriting (7.22) in the following form:

$$\gamma(x_k, \hat{x}_k) = \beta\bar{\gamma}, \quad \beta \in (0, 1). \tag{7.25}$$

Consequently, the Eq. (7.23) can be reduced to:

$$e_{k+1} = (A - K\bar{C}(\beta))e_k + W_1w_k - KW_2v_k. \tag{7.26}$$

In this equation, $\bar{C}(\beta) = [\beta\bar{\gamma}, \ 1, \ 1]$. This equation can be formulated in the following compact form:

$$e_{k+1} = X(\beta)e_k + Zz_k. \tag{7.27}$$

In this equation, $z_k = [w_k^T, v_k]^T$ and

$$X(\beta) = A - K\bar{C}(\beta), \quad Z = \bar{W}_1 - K\bar{W}_2, \quad \bar{W}_1 = W_1[I_{n_w} \; 0_{n_w \times 1}], \quad \bar{W}_2 = W_2[0_{1 \times n_w} \; 1]. \tag{7.28}$$

The system (7.27), which results from this, is a forced dynamic system with the input $z(k)$. Unfortunately, this property eliminates the possibility to directly employ the Lyapunov theory for the analysis of its convergence. In order to solve this problem, a Quadratic Boundedness (QB) approach [2] can be used, which has already been successfully applied in estimation and prediction schemes [36, 61].

For the sake of further analysis, it is concluded that the noise/disturbance vector z_k is bounded in the following manner:

$$z_1 \in \mathbb{E}_z, \quad \mathbb{E}_z = \{z : z^T Q z \le 1\}, \quad Q \succ 0. \tag{7.29}$$

This assumption enables the introduction of the QB paradigm and the following *Definition 1* [2, 11, 12] can be adapted to the system. For this system, the Lyapunov function is defined by $V_l = e_l^T P e_l$, $P \succ 0$:

Definition 1 The system (7.27) is strictly quadratically bounded for all allowable $z \in \mathbb{E}_z$, if $V_1 > 1$ implies $V_{k+1} < V_1$ for any $z \in \mathbb{E}_z$.

It should be noted that the strict quadratic boundedness of (7.27) guarantees that $V_{k+1} < V_k$ for any $z \in \mathbb{E}_z$ when $V_1 > 1$.

This assumption and definition enable the formulation of the central result of this section given by the subsequent theorem:

Theorem 7.1 *The system (5.45) is strictly quadratically bounded for all E_k and all allowable $\bar{w}_k \in \mathscr{E}_w$, if $N, P \succ 0$ and $0 < \alpha < 1$ exist such that the following conditions can be satisfied: The estimator (7.18) is convergent in the quadratically bounded sense, if $\alpha \in (0, 1)$, $P \succ 0$ and N exist such that the subsequent inequality is satisfied for all $\beta \in (0, 1)$:*

$$\begin{bmatrix} -P - \alpha P & 0 & A^T P - \bar{(C)}(\beta)^T N^T \\ 0 & -\alpha Q & \bar{W}_1^T P - \bar{W}_2^T N^T \\ PA - N\bar{(C)}(\beta) & P\bar{W}_1 - N\bar{W}_2 & -P \end{bmatrix} \prec 0. \tag{7.30}$$

Proof Using *Definition 1* system (7.27) is strictly quadratically bounded, if $P \succ 0$ and a scalar $\alpha \in (0, 1)$ exist such that:

$$\begin{bmatrix} X(\beta)^T P X(\beta) - P + \alpha P & X(\beta)^T P Z \\ Z^T P X(\beta) & Z^T P Z - \alpha Q \end{bmatrix} \prec 0, \tag{7.31}$$

for all $\beta \in (0, 1)$. Consequently, by employing the Schur complements (i.e. a a matrix that is calculated from the individual blocks of a larger matrix) to (7.31) and then multiplying both sides by diag (I, I, P) and substituting

$$N = PK, \tag{7.32}$$

give (7.30), which completes the proof. $\qquad\qquad\qquad\qquad\qquad\qquad\qquad\square$

Unfortunately, it is impossible to solve (7.30) directly for any $\beta \in (0, 1)$. Nevertheless, considering the linearity of $\bar{C}(\beta)$ with respect to $\beta \in (0, 1)$, two constants β_1 and β_2 can be defined that are close enough to either 0 or 1. This fact implies that these constants should satisfy the conditions $\beta_1 > 0$ and $\beta_2 < 1$. Consequently, instead of solving (7.30) it is corresponding to solve:

$$\begin{bmatrix} -P - \alpha P & 0 & A^T P - \bar{(C)}(\beta_i)^T N^T \\ 0 & -\alpha Q & \bar{W}_1^T P - \bar{W}_2^T N^T \\ PA - N\bar{(C)}(\beta_i) & P\bar{W}_1 - N\bar{W}_2 & -P \end{bmatrix} \prec 0, \quad i = 1, 2. \tag{7.33}$$

In conclusion, the design procedure of the innovative estimator can be realised in the subsequent form:

Off-line:

1. Obtain the upper bound $\bar{\gamma}$,
2. Select the overbounding matrix $Q \succ 0$ in (7.29),
3. Select $\alpha \in (0, 1)$,
4. Solve the "Linear Matrix Inequalities" (LMIs) (7.33) and obtain the gain matrix $K = P^{-1}N$.

On-line

5. Set $\hat{x}_0$ and $l = 0$.
6. Obtain the state estimate $\hat{x}_{k+1}$ according to (7.18).
7. Set $l = l + 1$ and go to *Step 2*.

It is an advantageous quality that $\alpha \in (0, 1)$ can be employed for controlling the convergence rate of the estimator, i.e. the larger α is chosen the faster will be the convergence; this appealing fact has been established in [36]. Consequently, (7.33) should be solved simultaneously with an iteratively modification of $\alpha \in (0, 1)$.

Based on the precedent elaboration, a states estimate $\hat{x}_k$ and, especially, $\hat{x}_{1,k}$, which is an estimate of the current state of charge (SOC) of the battery, can be generated. Consequently, the future behaviour of the battery can be predicted – this assignment is the focus of the next section.

7.6 Prognosis of the Remaining Useful Life

In this chapter, the state of charge (SOC) of the battery of an AGV represents the central measure of its health-aware performance. This proposition is motivated by the fact that the batteries of both AGVs should be able to supply a whole production day (two shifts). Both alternative scenarios – changing the battery or recharging the battery – are connected with considerable economical and technical disadvantages. Consequently, the central task of the following elaborations is the development a reliable predictor of the remaining SOC.

Considering that the description (7.16) is linear, a straight-forward concept is modelling this description linearly employing the following equation:

$$x_{1,l,i} = a_i l \Delta T + b_i, \quad i = 1, 2. \tag{7.34}$$

In this equation, a_i and b_i denote unknown parameters, which shape the relation between the time $l \Delta T$ and the state of charge (SOC) $x_{1,l}$ for the battery of the ith AGV. considering the challenging fact that the actual SOC is not available, the precedent equation can be replaced employing the subsequent equation:

$$\hat{x}_{1,l,i} = a_i l \Delta T + b_i, \quad i = 1, 2. \tag{7.35}$$

In this equation, $\hat{x}_{1,l,i}$ denotes an estimate of the SOC which can be obtained employing the innovative estimator which has been developed in Sect. 7.5.

In order to solve the estimation problem of a_i and b_i, a celebrated Recursive Least Square (RLS) algorithm can be applied:

$$\hat{p}_{l,i} = \hat{p}_{l-1,i} + K_{l,i}(\hat{x}_{1,l,i} - r_{l,i}^T \hat{p}_{l-1,1}), \tag{7.36}$$

$$K_{l,i} = P_{l-1,i} r_{l,i} \left(1 + r_{l,i}^T P_{l-1,i} r_{l,i} \right)^{-1}, \tag{7.37}$$

$$P_{l,i} = [I_2 - K_{l,i} r_{l,i}^T] P_{l-1,i}, \tag{7.38}$$

where the parameter estimation vector along with its regressor are defined by:

$$\hat{p}_{l,i} = [\hat{a}_{l,i}, \hat{d}_{l,i}]^T, \tag{7.39}$$

$$r_{l,i} = [l \Delta T, 1]^T. \tag{7.40}$$

To conclude, for obtaining a_i and b_i the subsequent algorithm can be applied:

Algorithm 1: Estimation algorithm

- Set $\hat{p}_{0,i} = [0, 1]^T$, $P_{0,1} = \delta I_2$ and $l = 0$, with $\delta > 0$ being a sufficiently large positive constant.
- Obtain the parameter estimates employing (7.36)–(7.38).
- Set $l = l + 1$ and go to Step 2.

As a consequence of the observation of (7.35) with the initial estimate $\hat{p}_{0,i} = [\hat{a}_{0,i}, \hat{d}_{0,i}]^T = [0, 1]^T$, it is evident that it associates to a state of charge (SOC) of the AGV battery equal to one. Actually, it is a natural assumption to conclude that the AGV batteries are completely charged at the beginning of the operation of these AGVs.

To conclude, by combining the algorithm of estimating the state of charge (presented in Sect. 7.5) with the one describing the estimation of the unknown parameters of (7.35), the subsequent SOC predictor can be generated:

$$\bar{x}_{1,l_x,i} = \hat{a}_{i,l}l_x\Delta T + \hat{d}_{i,l}, \quad i = 1, 2. \tag{7.41}$$

In this predictor, l_x is the discrete time of prediction based on the parameter estimates generated up to l. This implies that $l_x \geq l$.

The performance of the SOC estimation and of SOC prediction strategies is evaluated in the following section.

7.7 Performance Evaluation

Today it is still common to employ lead-acid batteries for supplying AGVs in production environments, most probably because of their economic advantages in terms of investment cost. Nonetheless, lead-acid batteries dispose of a relatively poor capacity-to-weight and capacity-to-volume ratio and require long charging processes. For the AGV operation described in this chapter, batteries with lithium-ion-technology are appropriate. Suitable batteries employ, for instance, a special kind of lithium-Ion-technology – the lithium nickel manganese cobalt oxide technology (Li-NMC); this technology is employed in bike and automotive applications, as well. Central advantages are a high energy density and minor self-heating. The cell voltage of Li-NMC cells lays between 3, 6 and 3, 7 V. An appropriate battery pack, delivers a nominal voltage of 25, 2 V and disposes of a nominal capacity of 200 Ah.

The first step of the evaluation is to provide numerical parameters related to the battery description (7.5), (7.6): $Q_{nom} = 200[Ah]$, $R_{CT} = 0.051[\Omega]$, $C_{CT} = 0.1922[F]$, $R_{Dif} = 0.0126[\Omega]$, $C_{Dif} = 0.8213[F]$, $R_0 = 0.05[\Omega]$, $\Delta t = 0.1[s]$.

On the basis of these parameters, the function $h(\cdot)$ of the battery pack needs to be determined. For this endeavour the general approach proposed in [57] along with a suitable polynomial regression function can be applied.

The function $h(\cdot)$ is illustrated in Fig. 7.16.

Figure 7.17 presents the function $\gamma(\cdot)$.

Based on this preliminary information, the *off-line* phase of the algorithm, which was explained in Sect. 7.5, can be started. Actually, the upper bound of $\gamma(\cdot)$ can be found without large efforts; and the upper bound of $\gamma(\cdot)$ is equal to $\bar{\gamma} = 45$. The following measure was to select the matrix $Q > 0$, which is shaping the overbounding ellipsoid of w_k and v_k. Considering the quantization error of the measurement unit

Fig. 7.16 Illustration of function $h(\cdot)$

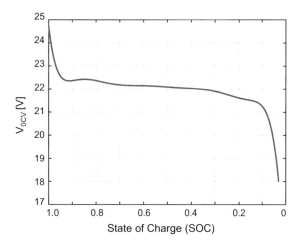

Fig. 7.17 Illustration of function $\gamma(\cdot)$

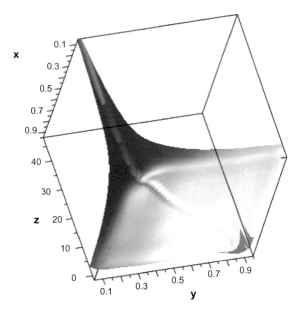

together with possible alterations of the internal capacities and/or resistances of the battery, it was selected as $Q = 100 I_{n_w+1}$. The following measure was to select the parameter $\alpha \in (0, 1)$, which is shaping the convergence of the estimator. The highest possible value was chosen $\alpha = 0.99$ and as a result of solving the LMIs (7.33) the estimator gain matrix could be obtained $K = [0.0225, 0.0001, 0.0011]^T$.

Based on the precedent parameter, the initial estimate was set to $\hat{x}_0 = [0.75, 0, 0]^T$. The simulation reproduced a usual manoeuvre of an AGV, which contains driving and stopping phases without an additional load. No alternate charging or discharging was carried out in order to avoid erroneous actual SOC predictions. The results of

Fig. 7.18 Actual SOC and its estimate – initial estimation phase (black solid: actual; red dash-dot: estimated)

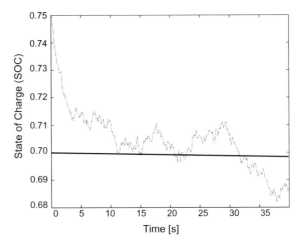

Fig. 7.19 Actual SOC and its estimate (black: actual; red: estimated)

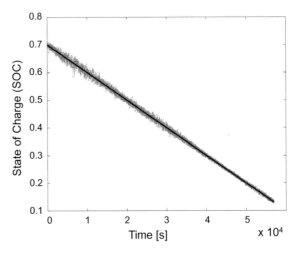

the investigation are illustrated in Fig. 7.18; in this investigation, the initial actual SOC was at the level of 0.7 and $\hat{x}_{1,0} = 0.75$ was selected intentionally to illustrate the convergence of the proposed estimator.

Figure 7.18 shows the actual SOC and its estimate in the initial estimation phase. A longer period of time is illustrated in Fig. 7.19.

The results shown in Figs. 7.18 and 7.19 clearly illustrate that the estimated SOC converges towards the actual SOC.

These results allow a recommendation of the proposed estimator for predicting the SOC according to the approach proposed in Sect. 7.6. For this scenario a typical working sequence of a AGV was created, which contains two alternate stages: 2 min transportation task with load and 2 min (idle) drive without load. In the transportation stage the average current is typically increasing by approx. 47% due to the load.

Fig. 7.20 SOC and its
prediction (black solid:
actual; red dash-dot:
predicted)

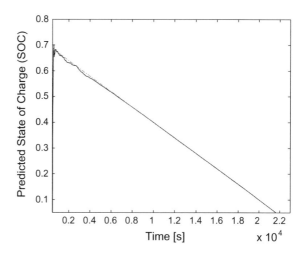

The assignment was to predict the SOC after 7 min and 30 s from each l; this indicates that l_x in (7.41) was set to $l_x = l + 4500$. The resulting predictions are shown in Fig. 7.20 along with the actual values.

In this outcome it is clearly visible that the uncertainty of the initial SOC value, which was set to 0.85 instead of the actual value equal to 0.8, leads to a large prediction error in the initial phase. The same kind of behaviour can also be observed in Fig. 7.18. After a transient phase, the prediction quality performance is significantly enhanced. This outcome leads to the recommendation to apply the developed SOC estimation/prediction scheme within both AGVs; this point is elucidated in the next section of this chapter.

7.8 Health-aware Model Predictive Control of the Assembly System

A model predictive control-based fault-tolerant strategy has been developed in earlier scientific activities (Chap. 6). This strategy has been expanded to systems with concurrency and is described in detail in [37]). This section describes the expansion to a health-aware model predictive control.

For assessing the useful operation potential of an AGV battery both its state of charge (SOC) and its state of health (SOH) have to be considered; this simultaneous presence is a major challenge for the planning and control of the AGV operations. In the following part of this section a second definition is presented, which allows to connect both aspects:

Definition 2 The state of health of a battery SOH is a number of cycles k_f from its full state of charge SOC equal to 1 down to zero charge.

Fig. 7.21 Illustration of the state of health of batteries

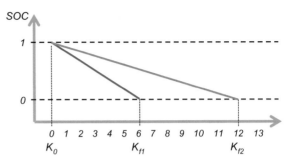

Figure 7.21 illustrates this definition; in this figure k_{f1} represents the feasible number of cycles of AGV_1 while k_{f2} represents the remaining number of cycles.

It can be clearly seen in Fig. 7.21 that the state of health of the first AGV (AGV_1) is definitely lower than the state of health of the second AGV (AGV_2). From the equation for the prediction of the SOC (7.41), it can be deduced that by means of substituting $l_x \Delta t = k_{f,i}(f_i(k_0) + b_i(k_0) + l\Delta t)$, the number of cycles $k_{f,i}$ to zero SOC follow the subsequent equation:

$$0 = \hat{a}_{i,l}(k_{f,i}(f_i(k_0) + b_i(k_0) + l\Delta t) + \hat{d}_{i,l} \tag{7.42}$$

It is important to note that the forward and backward drive times $f_i(k_0) + b_i(k_0)$ are employed for the purpose of prediction, i.e. it is assumed constant during performing the described prediction, i.e., $f_i(k) + b_i(k) = f_i(k_0) + b_i(k_0)$, for $k = k_0, \ldots, k_0 + N_p$.

The total number of forward and backward cycles of the ith AGV may be estimated using the subsequent equation:

$$k_{f,i} = \left\lfloor -\frac{\hat{a}_{i,l}l\Delta t + \hat{d}_{i,l}}{\hat{a}_i(f_i(k_0) + b_i(k_0))} \right\rfloor \tag{7.43}$$

In this equation, $\lfloor \cdot \rfloor$ rounds the resulting value to the smallest positive integer.

Consequently, for a given k_0 the remaining number of cycles is equal to

$$k_{r,i} = k_{f,1} - k_0 \tag{7.44}$$

Based on these consideration, a health-aware cost function for each of the two AGVs can be formulated. For the first AGV this function is:

$$J_{h,1} = (f_1(k_0) + b_1(k_0))k_{r,1} - (f_1(k_0) + b_1(k_0)) \sum_{k=k_0}^{k_0+N_p} (1 - z(k)). \tag{7.45}$$

In this function $(f_1(k_0) + b_1(k_0))k_{r,1}$ denotes the total operational time of the first AGV, whereas $(f_1(k_0) + b_1(k_0)) \sum_{k=k_0}^{N_p} (1 - z(k))$ stands for its exhausted portion.

For the second AGV this function is:

$$J_{h,2} = (f_2(k_0) + b_2(k_0))k_{r,2} - (f_2(k_0) + b_2(k_0)) \sum_{k=k_0}^{k_0+N_p} z(k) \qquad (7.46)$$

The cost functions of both AGVs can be added up to a total health-aware cost function using the following equation:

$$J_h = J_{h,1} + J_{h,2}. \qquad (7.47)$$

This cost function has to be maximised.

A last step can be the merging of the cost function resulting form the conventional MPC (compare [37]) with J_h; this can be realised using the following equation:

$$J = (1 - \kappa)J_y + \kappa J_h \qquad (7.48)$$

where $0 \le \kappa \le 1$.

In this total cost function the variable $\kappa = 0$ represents the importance of the SOC. If $\kappa = 0$, then the cost function does not consider the state of charge SOC and the state of health of the AGV batteries. On the contrary, if $\kappa = 1$, then the state of charge and the state of health are the only important optimizations. Consequently, the control engineer needs to find an optimum trade-off between these two situations by an appropriate selection of κ.

To conclude, it should be pointed out that two simultaneous algorithms form the proposed health aware control strategy. The first algorithm leads to estimation of the current values of $\hat{a}_{i,l}$ and $\hat{d}_{i,l}$ (compare Sect. 7.6). The second algorithm concerns the precedent predictive control and can be illustrated in the following form:

Algorithm 2: Predictive control algorithm

Step 0:   Set $N_p > 0$, $\kappa \in [0, 1]$, $x(0)$, $k = 0$.
Step 1:   Set $k_0 = k$ and calculate $k_{f,1}$ and $k_{f,2}$ according to (7.43).
Step 2:   Solve a mixed-integer linear programming problem

$$(y^*(k_0), \ldots y^*(k_0 + N_p)) = \arg \max_{\substack{x(k_0),\ldots,x(k_0+N_p) \\ y(k_0),\ldots,y(k_0+N_p) \\ z(k_0),\ldots,z(k_0+N_p)}} J, \qquad (7.49)$$

under the given constraints listed in [37].

Step 3:   Apply $y(k) = y^*(k_0)$ to the cooperative AGV system with $z(k_0)$ determining AGV performing transportation for kth event counter.

Step 4:   Set $k = k + 1$ and go to *Step 1*.

7.9 Verification and Experimental Results

The main objective of this section is to verify the performance of the proposed approach. In order to approach this objective, three different scenarios were evaluated:

Scenario 1:　This scenario describes a predictive control task without health-aware features ($\kappa = 0$).

Scenario 2:　This scenario describes a predictive control task with health-aware features; in this scenario the batteries of both AGVs dispose of the same SOC; the weighting factor was chosen as $\kappa = 0.5$.

Scenario 3:　This scenario describes a predictive control task with health-aware features, the SOC of the first AGV (AGV_1) is smaller than the SOC of the second AGV (AGV_2); the weighting factor was chosen as $\kappa = 0.5$.

In all three scenarios the initial condition $x(0) = 0$ was employed and the constraint concerning the schedule $t_{ref}(k)$ was formed by the subsequent sequence:

$$10, \ 13, \ 15, \ 18, \ 21, \ 23, \ 26, \ 29, \ 31, \ 34, \ 37, \ 39, \ \ldots \tag{7.50}$$

The precedent sequence deliberately exhibits an irregularity – in this manner an uncomplicated switching between the two cooperative AGVs could be avoided.

The optimum sequence of $y(k)$ and $x_3(k)$ which satisfies (7.50) and was obtained for *Scenario 1* is shown in Fig. 7.22.

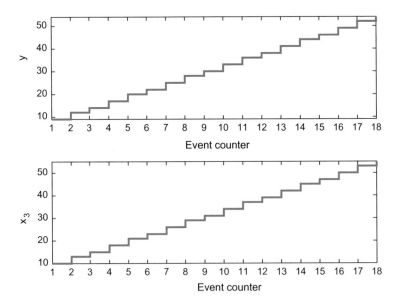

Fig. 7.22 Scenario 1: optimum sequence of $y(k)$ and $x_3(k)$

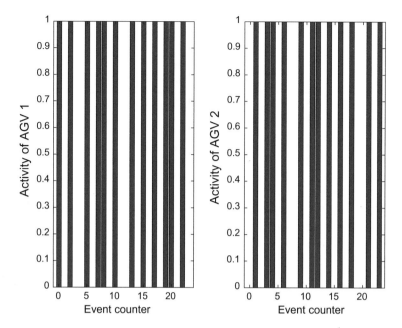

Fig. 7.23 Scenario 1 and Scenario 2: AGV activities

In this case, the activities of both AGVs are correspondingly spread during the time period of operation—this can be seen in Fig. 7.23.

Nevertheless, an easier schedule than (7.50) might lead to a loss of this balance and only one of the AGVs might be used exclusively. This might lead to a complete depletion of the battery of this AGV – at the same time the battery of the other AGV might still be full. It is possible to avoid this kind of situation, when the developed approach is employed.

As an intermediate step, *Scenario 2* disposes of health aware-features but in this scenario an equal SOH is expressed by the remaining cycles $k_{f,i}$. It is important to note that exactly the same optimum $y(k)$ and resulting $x_3(k)$ are generated for all three scenarios (compare Fig. 7.22). The distinction between the three scenarios can be found in the manner the AGVs are used. The activities for both AGVs for *Scenario 2* are exactly equal to those in *Scenario 1* and can consequently also be seen in Fig. 7.23. For a more detailed discussion, Fig. 7.24 contains the evolution of $k_{f,i}$ for *Scenario 2*.

In the case of *Scenario 2*, it is obvious that both are decreasing correspondingly. This behaviour is different in *Scenario 3* – just because one AGV battery disposes of a smaller $k_{f,i}$. The AGV activities are shown in Fig. 7.25

Again, for a detailed discussion the evolution of $k_{f,i}$ for *Scenario 3* is shown in Fig. 7.26.

The positive effects of the developed algorithm are instantly visible; no alternating behaviour of the AGVs can be identified any more. Here AGV_2 is employed for much

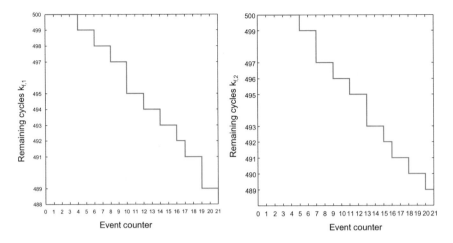

Fig. 7.24 Scenario 2: evolution of $k_{f,i}$

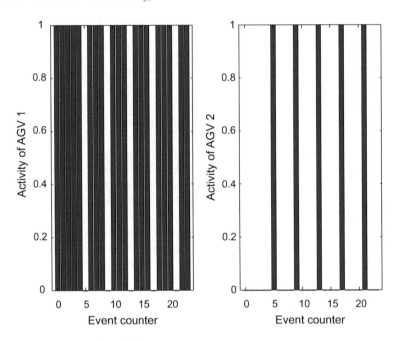

Fig. 7.25 Scenario 3: AGV activities

more activities and AGV_1 is only active, when it is necessary in order to meet the required schedule. It is obvious that the developed algorithm enables a balanced exploitation of the collaborating AGVs – such system behaviour may significantly enhance the capabilities of the cooperative system. It is important to point out that the developed algorithm guarantees that a optimum sequence is found.

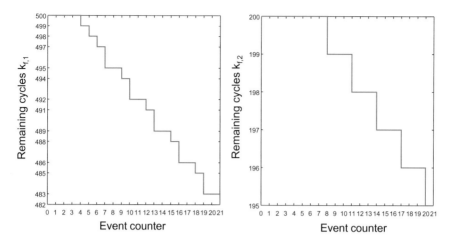

Fig. 7.26 Scenario 3: evolution of $k_{f,i}$

Please note that the transportation times and processing times are obscured and simplified.

7.10   Conclusions

In the producing industry, the flexibility of manufacturing and assembly systems has become a key success factor, because an agile reaction is mandatory in competitive industrial sectors. This is especially true in the current automotive industry which is characterised by a profound transition caused by autonomous driving and electro mobility. In the logistic processes within the plants of automotive industry, AGVs have excelled because of their operational potential and flexibility. However, their advantages can only be exploited to their full extent, if reliable control systems are present which can be applied in complex scenarios and which can tolerate certain unavoidable faults. The main contribution of this chapter is a framework for health aware model predictive control. This framework allows to take aspects of the health status and the remaining useful life of an AGV battery into consideration when the activities of cooperative, redundant AGVs have to be planned and controlled. The framework will enable either a relaxation of the schedules in a manufacturing or assembly system or to balance the RUL of the batteries of the AGVs. A sample application in a seat assembly plant shows the application potential. So far, recharge possibilities have not been considered; this might be an interesting direction for further research.

References

1. Adams, D.: Health Monitoring of Structural Materials and Components: Methods with Applications. Wiley-Interscience, New Jersey (2007)
2. Alessandri, A., Baglietto, M., Battistelli, G.: Design of state estimators for uncertain linear systems using quadratic boundedness. Automatica **42**(3), 497–502 (2006)
3. Amstadter, B.L.: Reliability Mathematics: Fundamentals, Practices. Procedures. McGraw-Hill, New York (1977)
4. Andre, D., Appel, C., Soczka-Guth, T., Sauer, D.U.: Advanced mathematical methods of soc and soh estimation for lithium-ion batteries. J. Power Sour. **224**, 20–27 (2013)
5. Baccelli, F., Cohen, G., Olsder, G.J., Quadrat, J.P.: Synchronization and linearity: an algebra for discrete event systems. J. Oper. Res. Soc. **45**, 118–119 (1994)
6. Barre, A., Deguilhem, B., Grolleau, S., Gerad, M., Suard, F., Riu, D.: A review on lithium-ion battery ageing mechanisms and estimations for automotive applications. J. Power Sour. **241**, 680–689 (2013)
7. Berecibar, M., Gandiaga, I., Villarreal, I., Omar, N., Van Mierlo, J., Van den Bossche, P.: Critical review of state of health estimation methods of li-ion batteries for real applications. Renew. Sustain. Energy Rev. **56**, 572–587 (2016)
8. Butkovic, P.: Max-Linear Systems: Theory and Algorithms. Springer, Berlin (2010)
9. Chen, Z., Cao, M., Mao, Z.: Remaining useful life estimation of aircraft engines using a modified similarity and supporting vector machine (svm) approach. Energies **11**(1), (2018)
10. Dabrowska, A., Stetter, R., Sasmito, H., Kleinmann, S.: Extended kalman filter algorithm for advanced diagnosis of positive displacement pumps. In: Proceedings of the 8th SAFEPROCESS: IFAC International Symposium on Fault Detection, Supervision and Safety for Technical Processes, 29th to 31st August 2012. Mexico City, Mexico (2012)
11. Ding, B.: Constrained robust model predictive control via parameter-dependent dynamic output feedback. Automatica **46**(9), 1517–1523 (2010)
12. Ding, B.: Dynamic output feedback predictive control for nonlinear systems represented by a Takagi-Sugeno model. IEEE Trans. Fuzzy Syst. **19**(5), 831–843 (2011)
13. Dong, M., He, D.: A segmental hidden semi-markov model (hsmm)-based diagnostics and prognostics framework and methodology. Mech. Syst. Signal Process. **21**, 2248–2266 (2007)
14. Farrar, C.R., Worden, K.: An introduction to structural health monitoring. Philos. Trans. R. Soc. (Math., Phys. Eng. Sci.) **365**(1), 303–315 (2007)
15. Ferri, F.A.S., Rodrigues, L.R., Gomes, J.P.P., Medeiros, I.P., Galvao, R.K.H., Nascimento Jr C.L.: Combining phm information and system architecture to support aircraft maintenance planning. In: Proceedings of the IEEE International Systems Conference, Orlando, USA (2013)
16. Goebel, K., Saha, B., Saxena, A., Celaya, J.R., Christophersen, J.: Prognostics in battery health management. IEEE Instrum. Meas. Mag. **11**(4), 33–40 (2008)
17. He, W., Willard, N., Osterman, M., Pecht, M.: Prognostics of lithium-ion batteries based on dempstershafer theory and the bayesian monte carlo method. J. Power Sour. **196**, 10314–10321 (2011)
18. Heng, A., Zhang, S., Tan, A.C.C., Mathew, J.: Rotatin gmachinery prognostics: state of the art, challenges and opportunities. Mech. Syst. Signal Process. **23**, 724–739 (2009)
19. Holder, K., Zech, A., Ramsaier, M., Stetter, R., Niedermeier, H.-P., Rudolph, S., Till, M.: Model-based requirements management in gear systems design based on graph-based design languages. Appl. Sci. **7**, (2017)
20. Hu, C., Youn, B.D., Chung, J.: A multiscale framework with extended kalman filter for lithium-ion battery soc and capacity estimation. Appl. Energy **92**, 694–704 (2012)
21. Hua, Y., Cordoba-Arenas, A., Warner, N., Rizzoni, G.: A multi time-scale state-of-charge and state-of-health estimation framework using nonlinear predictive filter for lithium-ion battery pack with passive balance control. J. Power Sour. **280**, 293–312 (2015)
22. Huang, R., Xi, L., Li, X., Liu, C.R., Qiu, H., Lee, J.: Residual life predictions for ball bearings based on self-organizing map and back propagation neural network methods. Mech. Syst. Signal Process. **21**, 193–207 (2007)

23. Isermann, R.: Fault Diagnosis Systems. An Introduction from Fault Detection to Fault Tolerance. Springer, New York (2006)
24. Kan, M.S., Tan, A.C.C., Mathew, J.: A review on prognostic techniques for non-stationary and non-linear rotating systems. Mech. Syst. Signal Process. **62–63**, 1–20 (2015)
25. Kececioglu, D.: Reliability Engineering Handbook, vol. 2. Wiley-Interscience, New Jersey (2002)
26. Khorasgani, H., Biswas, G., Sankararaman, S.: Methodologies for system-level remaining useful life prediction. Reliab. Eng. Syst. Saf. **154**, 8–18 (2016)
27. Klass, V., Behm, M., Lindbergh, G.: A support vector machine-based state-of-health estimation method for lithium-ion batteries under electric vehicle operation. J. Power Sour. **270**, 262–272 (2014)
28. Klutke, G.-A., Kiessler, P.C., Wortman, M.A.: A critical look at the bathtub curve. IEEE Trans. Reliab. **52**(1), 125–129 (2003)
29. Kodagoda, K.R.S., Wijesoma, W.S., Teoh, E.K.: Fuzzy speed and steering control of an agv. IEEE Trans. Control. Syst. Technol. **10**(1), 112–120 (2002)
30. Lee, J., Wu, F., Zhao, W., Ghaffari, M., Liao, L., Siegel, D.: Prognostics and health management design for rotary machinery systems - reviews, methodology and applications. Mech. Syst. Signal Process. **42**, 314–334 (2014)
31. Lei, Y., Li, N., Guo, L., Li, N., Yan, T., Lin, J.: Machinery health prognostics: a systematic review from data acquisition to rul prediction. Mech. Syst. Signal Process. **104**, 799–834 (2018)
32. Li, N., Lei, Y., Liu, Z., Lin, J.: A particle filtering-based approach for remaining useful life predication of rolling element bearings. In: 2014 International Conference on Prognostics and Health Management, pp. 1–8 (2014)
33. Liao, L., Koetting, F.: Review of hybrid prognostics approaches for remaining useful life prediction of engineered systems, and an application to battery life prediction. IEEE Trans. Reliab. **63**(1), 191–207 (2014)
34. Majdzik, P., Akielaszek-Witczak, A., Seybold, L., Stetter, R., Mrugalska, B.: A fault-tolerant approach to the control of a battery assembly system. Control Eng. Pract. **55**, 139–148 (2016)
35. Mosallam, A., Medjaher, K., Zerhouni, N.: Data-driven prognostic method based on bayesian approaches fordirect remaining useful life prediction. J. Intell. Manuf. **27**(5), 1037–1048 (2016). Oct
36. Mrugalska, B.: A bounded-error approach to actuator fault diagnosis and remaining useful life prognosis of takagi-sugeno fuzzy systems. ISA Trans. **80**, 257–266 (2018)
37. Mrugalska, B., Stetter, R.: Health-aware model-predictive control of a cooperative AGV-based production system. Sens. **19**(3), (2019)
38. Nuhic, A., Terzimehic, T., Soczka-Guth, T., Buchholz, M.: Health diagnosis and remaining useful life prognostics of lithium-ion batteries using data-driven methods. J. Power Sour. **239**, 680–688 (2013)
39. Paoli, A., Sartini, M., Lafortune, S.: Active fault tolerant control of discrete event systems using online diagnostics. Automatica **47**, 639–649 (2011)
40. Pecht, M.: Prognostics and Health Management of Electronics. Wiley-Interscience, New York (2010)
41. Pecht, M., Jaai, R.: A prognostics and health management roadmap for information and electronics-rich systems. Microelectron. Reliab. **50**, 317–323 (2010)
42. Peng, Y., Dong, M., Zuo, M.J.: Current status of machine prognostics in condition-based maintenance: a review. Int. J. Adv. Manuf. Technol. **50**(1–4), 297–313 (2010)
43. Pham, H.T., Yang, B.S.: Estimation and forecasting of machine health condition using arma/garch model. Mech. Syst. Signal Process. **24**(2), 546–558 (2010)
44. Polak, M., Majdzik, Z., Banaszak, P., Wojcik, R.: The performance evaluation tool for automated prototyping of concurrent cyclic processes. Fundam. Inform. **60**, 269–289 (2004)
45. Pordeus Gomes, J.P., Rodrigues, L.R., Harrop Galvao, R.K., Yoneyama, T.: System level rul estimation for multiple-component systems. In: Proceedings of the 1st Annual Conference of the Prognostics and Health Management Society (2013)

46. Rad, J.S., Hosseini, E., Zhang, Y., Chen, C.: Online tool wear monitoring and estimation using power signals and s-transform. In: Proceedings of SysTol (2013)
47. Ramsaier, M., Holder, K., Zech, A., Stetter, R., Rudolph, S., Till, M.: Digital representation of product functions in multicopter design. In: Proceedings of the 21st International Conference on Engineering Design (ICED 17) Vol 1: Resource Sensitive Design, Design Research Applications and Case Studies (2017)
48. Remmlinger, J., Buchholz, M., Soczka-Guth, T., Dietmayer, K.: On-board state-of-health monitoring of lithium-ion batteries using linear parameter-varying models. J. Power Sour. **239**, 689–695 (2013)
49. Rezvanizaniani, S.M., Liu, Z., Chen, Y., Lee, J.: Review and recent advances in battery health monitoring and prognostics technologies for electric vehicle (ev) safety and mobility. J. Power Sour. **256**, 110–124 (2014)
50. Saha, B., Goebel, K., Poll, S., Christophersen, J.: Prognostics methods for battery health monitoring using a bayesian framework. IEEE Trans. Instrum. Meas. **58**(2), 291–296 (2009)
51. Sankararaman, S.: Significance, interpretation, and quantification of uncertainty in prognostics and remaining useful life prediction. Mech. Syst. Signal Process. **52–53**, 228–247 (2015)
52. Seybold, L., Witczak, M., Majdzik, P., Stetter, R.: Towards robust predictive fault-tolerant control for a battery assembly system. Int. J. Appl. Math. Comput. Sci. **25**(4), 849–862 (2015)
53. Si, X.-S., Wang, W., Hu, C.-H., Zhou, D.-H.: Remaining useful life estimationa review on the statistical data driven approaches. Eur. J. Oper. Res. **213**(1), 1–14 (2011)
54. Sidhu, A., Izadian, A., Anwar, S.: Adaptive nonlinear model-based fault diagnosis of li-ion batteries. IEEE Trans. Ind. Electron. **62**(2), 1002–1011 (2015)
55. Sikorska, J.Z., Hodkiewicz, M., Ma, L.: Prognostic modelling options forremaining use ful life estimation by industry. Mech. Syst. Signal Process. **25**, 1803–1836 (2011)
56. Singleton, R.K., Strangas, E.G., Aviyente, S.: Extended kalman filtering for remaining-useful-life estimation of bearings. IEEE Trans. Ind. Electron. **62**(3), 1781–1790 (2015). March
57. Snihir, I., Rey, W., Verbitskiy, E., Belfadhel-Ayeb, A., Notten, P.: Battery open-circuit voltage estimation by a method of statistical analysis. J. Power Sour. **159**(2), 1484–1487 (2006)
58. Stetter, R., Paczynski, A.: Intelligent steering system for electrical power trains. In: Emobility Electrical Power Train - IEEEXplore, pp. 1–6 (2010)
59. Stetter, R., Paczynski, A., Zajac, M.: Methodical development of innovative robot drives. In: Tools and Methods of Competitive Engineering – TMCE 2008: Proceedings of the seventh international symposium. Izmir, Turcja, 2008. Delft: Delft University of Technology, vol. 1, pp. 565–576 (2008)
60. Stetter, R., Witczak, M.: Degradation modelling for health monitoring systems. J. Phys. **570**, (2014)
61. Stetter, R., Witczak, M., Pazera, M.: Virtual diagnostic sensors design for an automated guided vehicle. Appl. Sci. **8**(5) (2018)
62. Taborelli, C., Onori, S.: Advanced battery management system design for soc/soh estimation for e-bikes applications. Int. J. Powertrains **5**(4) (2016)
63. Tobon-Mejia, D.A., Medjaher, K., Zerhouni, N., Tripot, G.: A data-driven failure prognostics method based on mixture of gaussians hidden markov models. IEEE Trans. Reliab. **61**(2), 491–503 (2012)
64. van den Boom, T.J.J., De Schutter, B.: Modelling and control of discrete event systems using switching max-plus-linear systems. Control. Eng. Pract. **14**, 1199–1211 (2006)
65. Wang, H.-K., Li, Y.-F., Huang, H.-Z., Jin, T.: Near-extreme system condition and near-extreme remaining useful time for a group of products. Reliab. Eng. Syst. Saf. **162**, 103–110 (2017)
66. Wang, J., Steiber, J., Surampudi, B.: Autonomous ground vehicle control system for high-speed and safe operation. In: 2008 American Control Conference (2008)
67. Wang, T., Yu, J., Siegel, D., Lee, J.: A similarity-based prognostics approach for remaining useful life estimation of engineered systems. In: Proceedings of the International Conference on Prognostics and Health Management, pp. 1–6 (2008)
68. Widodo, A., Shim, M.-C., Caesarendra, W., Yang, B.-S.: Intelligent prognostics for battery health monitoring based on sample entropy. Expert. Syst. Appl. **38**, 11763–11769 (2011)

69. Witczak, M.: Fault Diagnosis and Fault-Tolerant Control Strategies for Non-Linear Systems. Analytical and Soft Computing Approaches. Springer, Berlin (2014)

70. Yousfi, Basma: Rassi, Tarek, Amairi, Messaoud, Aoun, Mohamed: Set-membership methodology for model-based prognosis. ISA Trans. **66**, 216–225 (2017)

71. Zhang, H., Hu, C., Kong, X., Zhang, W., Zhang, Z.: Online updating with a wiener-process-based prediction model using ukf algorithm for remaining useful life estimation. In: 2014 Prognostics and System Health Management Conference (PHM-2014 Hunan), pp. 305–309 (2014)

72. Ziemniak, P., Stania, M., Stetter, R.: Mechatronics engineering on the example of an innovative production vehicle. In: Norell Bergendahl, M., Grimheden, M., Leifer, L., Skogstad, P., Lindemann, U. (eds.) Proceedings of the 17th International Conference on Engineering Design (ICED'09), vol. 1. pp. 61–72 (2009)

73. Zou, Y., Hu, X., Ma, H., Li. S.E.: Combined state of charge and state of health estimation over lithium-ion battery cell cycle lifespan for electric vehicles. J. Power Sour. **270**, 793–803 (2015)

Chapter 8
Extension to Automated Processes with Flexible Redundant and Shared Elements

The last two chapters have focused on a strategy and framework for the predictive fault-tolerant control of complex automated processes. This section expands these topics towards automated processes which can dispose of both flexible redundant and shared elements. "Redundant elements" are elements within an automated process which are duplicated or multiplied in order to achieve certain advantages which are explained later in this section, such as an increase of reliability. It will also be discussed later in this section that it is advantageous to combine such elements with certain aspects of flexibility in order to achieve *flexible redundant* elements. "Shared elements" or "shared resources" are elements within an automated process which can be used in more than one process chain such as testing stations which perform the tests for two or more assembly lines. These shared elements are frequently used in current automated process because of, amongst others, economical reasons. This chapter describes an interval max-plus algebra fault-tolerant control framework, which can deal with automated processes containing flexible redundant and shared elements. Additionally, this framework addresses the problem of uncertainties, which are very common in the current industrial reality. Several recent investigations have been exposing considerable advantages of max-plus algebra in the context of uncertainty models [4, 7, 11]. The contents of this chapter are based on the publication [8] as well as the earlier works described in [5, 6, 10]. The important preliminaries concerning interval max-plus algebra have already been described in Sect. 6.2.1 of this book and will also be used as a basis in this chapter.

This chapter is organised as follows: flexible redundant and shared elements and their specific advantages are explained in Sect. 8.1. Section 8.2 describes the car seat assembly system being investigated. Section 8.3 concentrates on the modelling of redundant "Automated Guided Vehicles" AGVs, while Sect. 8.4 describes the assembly system modelling. In Sect. 8.5 the constrained model predictive control algorithm is described and in Sect. 8.6 the fault-tolerant control scheme is elucidated. The implementation results of these algorithms and schemes are discussed in Sect. 8.7.

© Springer Nature Switzerland AG 2020
R. Stetter, *Fault-Tolerant Design and Control of Automated Vehicles and Processes*, Studies in Systems, Decision and Control 201, https://doi.org/10.1007/978-3-030-12846-3_8

8.1 Flexible Redundant and Shared Elements

As stated in Sect. 3.4, redundancies, i.e. the duplication or multiplication of process elements, can be important means for fault-tolerant design. In automated processes, the realisation of redundancies can lead to several advantages:

- *Increasing the fault-tolerance and, through this, the reliability of the superordinate system*: Redundant elements can replace each other in the case of a fault and can allow to, at least partly, compensate the influence on a superordinate system and thus increase its reliability. For instance, in the case of AGVs in an assembly system, a second AGV can replace a faulty first AGV and can allow the continuation of the assembly sequence in the assembly system (the superordinate system).
- *Allowing scalability*: Scalable assembly systems can react to changes concerning the number of assembled goods; redundant elements can be used to realise this scalability. Changes in the number of assembled goods are very common in industry, because usually during product ramp-up the number of assembled goods is rather low, then it will reach its maximum and is decreasing again towards the "End Of Life" (EOL) of a product. These changes can also be a result of changing product demand depending on the season of the year and also economical fluctuations. An example for redundant elements in an automated process could be to use only one AGV for one task in the ramp-up phase, then expand to three redundant AGVs in the peak period and maybe two redundant AGVs close to the EOL of the assembled product.
- *Improving maintainability*: Redundant elements can improve maintainability, because one redundant element can replace other elements during their maintenance.
- *Increasing flexibility*: Redundant elements can contribute to increase the flexibility of superordinate systems, because they can buffer irregularities in the remaining process elements.

It is even easier and more probable to realise these advantages, if the redundant elements dispose of certain aspects of flexibility. The two most important aspects are *position flexibility* and *time flexibility*.

- *Position flexibility* can be achieved e.g. by wheels which allow an element to move free from rails, conveyors or similar fastening or guiding systems. The position flexibility of AGVs, which is one of their main advantages, allows that two or more redundant AGVs can use the same travel paths and leads to low space requirements and investment costs.
- *Time flexibility* refers to the capability of process elements to perform certain tasks or to arrive at certain positions independently from an predetermined schedule. Intelligent algorithms, elaborate control systems as well as secure and fast communication are required in order to allow a coordination of a superordinate system in spite of time flexibility; the science community has accepted this challenge and has proposed, developed and validated such algorithms and systems. Consequently, flexible redundant elements have found frequent application in

flexible redundant elements

position flexibility
- independent from fastening or guiding elements

time flexibility
- independent from predetermined schedule

application flexibility
- geometrical and material
- physical / functional

condition flexibility
- insensible to changing environment

Fig. 8.1 Aspects concerning flexible redundant elements

current and future automated processes. Scientific investigations show that it is possible to realise the control of redundant elements as elements of a system with concurrency and to apply switching max-plus algebra [13].

Two other aspects of flexibility also play a notable role in automated processes: *application flexibility* and *condition flexibility*:

- *Application flexibility*: this term describes the capability of process elements to be applied for more than one application. It is sensible to distinguish three levels of application flexibility: geometrical and material application flexibility, physical application flexibility and functional application flexibility (these three levels correspond to the levels described in Sect. 3). *Geometrical and material application flexibility* could be that an AGV is able to transport objects which differ in terms of geometry or material. *Physical application flexibility* could be that a AGV could realise the transportation by means of friction (e.g. a conventional gripper) or by means of negative pressure (e.g. a vacuum gripper). *Functional application flexibility* could be that an AGV can perform two different functions, e.g. to transport goods and to perform an assembly operation on these or other goods.
- *Condition flexibility*: the meaning of this term is that process elements can be applied in situations with changing conditions such as changing temperatures or changing floor conditions.

The listed aspects concerning flexible redundant elements are summarised in Fig. 8.1.

It is interesting to note that flexible redundant elements are also present in the organs of biological systems.

Shared elements (often also referred to as shared resources) are used in more than one process chain of an automated process. The most prominent reason for shared elements is economical: commonly expensive elements with a rather short processing time are used as shared elements in order to save primarily investment costs but also operation costs, due to possibly reduced maintenance and personnel demand. Sometimes, a temporary application of shared resources is necessary and sensible because of the failure of a redundant or similar element. Unfortunately,

**shared elements
(shared resources)**

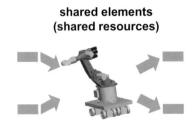

possible advantages:
- low investment costs
- efficient design
- light-weight design

challenges
- resource conflicts
- process bottlenecks
- reliability issues
  (superordinate system)

Fig. 8.2 Advantages and challenges concerning shared elements

the application of shared elements can lead to resource conflicts. In the case that the number of conflicts increases, the waiting time of all tasks in a process will increase. Consequently, the avoidance of resource conflicts is a prominent aspect to achieve a higher performance level of automated processes and their avoidance is in the centre of current research activities [3]. The usage of *shared elements* is in accordance with general design guidelines, because they can lead to more efficient and light-weight systems. However, from the viewpoint of fault-tolerant design, shared elements can be critical, because they can decrease the reliability of the overall system and can lead to conflicts and bottlenecks. This should not indicate that shared elements should be avoided, but they should be applied with special attention and should dispose of comparatively high reliability and of self-diagnosis or even predictive self-diagnosis capabilities. This is also a promising field for future research. Figure 8.2 gives an overview of the listed advantages and challenges concerning shared elements.

Flexible redundant and shared elements lead to additional specific constraints for automated processes which can be defined in the proposed control framework. This framework allows avoiding resource conflicts or at least to minimise their influence and can realise a smooth and effective cooperation between flexible redundant elements in automated processes.

8.2 Sample Process

The sample application of the chapter is a seat assembly system, similar to the one described in Chap. 7. In contrast to Chap. 7, a dedicated robot for assembling safety equipment is used as a *shared element*. In the assembly scenario two variants of novel seat frames are transported from two initial assembly stations to this dedicated robot and to two final assembly stations (Fig. 8.3).

The first process step can either be the assembly station for basic seat frames (resource R_1; respective operation duration $d_1(k)$) or the assembly station for complex seat frames (resource R_2 respective operation duration $d_2(k)$). Both kinds of seat frames will then be transported to the next assembly station which disposes of a dedicated industrial robot (the shared element). This industrial robot is capable of assembling the safety systems (resource R_3; respective operation duration

Fig. 8.3 Sample scenario: process depiction

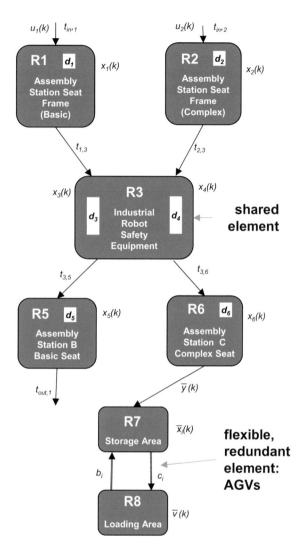

$d_3(k)$ for basic seat frames and operation duration $d_4(k)$ for complex seat frames). The industrial robot holds a screw fastening device which will measure and store certain information entities (screw torque etc.) required for quality and safety documentation. Basic seat frames are then transported to a first assembly station B (for Basic) complete seat (resource R_5; respective operation duration $d_5(k)$) and complex seat frames will be transported to the assembly station C (for Complex) complete seat (resource R_6; respective operation duration $d_6(k)$). This chapter concentrates on complex seat frames which are assembled in assembly station C (resource R_6). The complex seats are transferred to a storage area (R_7). From this storage area, two redundant AGVs (the flexible redundant elements) transport the completed complex seats to a loading area (R_8) located in another building. In this final stage, c_i denotes seat loading and transportation from the storage area R_7 to the loading area R_8 while b_i denotes seat unloading and transportation from the loading area R_8 to the storage area R_7. For the sake of consistency the operation durations d_7 and d_8 are introduced, however both of them are equal to zero, i.e. $d_7 = 0$ and $d_8 = 0$.

The automated process explained above is a flexible manufacturing system (FMS) and can sensibly be described as Discrete Event System (DES) with identifiable discrete states [9, 12, 14] using the following parameters:

- the set of processing operations $\mathbb{P} = \{p_{i,j}\}$, where:
  $p_{i,j}$ denotes the i-th processing operation performed on the j-th resource,
  $i \in \{1, 2, \dots, n\}$, n—the number of processing operations,
  $j \in \{1, 2, \dots, r\}$, r—the number of resources;
- the set of operation times $\mathbb{D} = \{d_1, d_2, \dots, n\}$;
- the set of operation times $\mathbb{T} = \{t_{l,k}\}$, $\forall l, k \in \{1, 2, \dots, r\} : l \neq k$ and there is a connection between l-th resource and k-th resource.

Based on this formal description of the process and the mathematical background of the interval max-plus algebra the mathematical model of redundant AGVs and the assembly system can be determined.

8.3   Modelling Flexible Redundant Elements

This section aims at describing a mathematical model of the flexible redundant elements—the redundant AGVs operating between storage and loading areas. The main objective is to generate a schedule realising the optimal performance of the AGVs; the first step of this endeavour is to describe their behaviour mathematically. In order to do so, the following variables are defined:

- $\bar{x}_i(k)$—transportation start time for i-th AGV for k-th event counter;
- $\bar{y}(k)$—storage zone seat availability time for k-th event counter;
- $\bar{v}(k)$—loading zone seat availability time for k-th event counter;
- $v_i(k)$—decision associated with i-th AGV for k-th event counter.

This set of variables allows to determine the desired mathematical model. One main challenge is the usage of flexible redundant elements (the AGVs) which can lead to concurrency (compare [13]). This challenge can be addressed by introducing decision variables $v_i(k)$ which can only assume values within a two-valued set $\{e, \varepsilon\}$. The meaning of this decision variables is that $v_i(k) = e$ describes that the i-th AGV performs a transportation task for k-th event counter while $v_i(k) = \varepsilon$ denotes an opposite situation. The i-th AGV can be described with the state equation:

$$\bar{x}_i(k + 1) = \max(\bar{x}_i(k) + b_i(k) + c_i(k), \bar{y}(k + 1)), \tag{8.1}$$

with the associated constraints:

$$v_i(k) = \varepsilon \quad \Leftrightarrow \quad b_i(k) = e, \; c_i(k) = e, \tag{8.2}$$

$$v_i(k) = e \quad \Leftrightarrow \quad v_j(k) = \varepsilon, \; \forall_{i \neq j}. \tag{8.3}$$

The meaning of constraint 8.2 concerns the situation that the i-th AGV is not realising a transportation task in this cycle with the k-th event counter; in this case the transportation times in both directions are equal to zero. The second constraint guarantees that exclusively the i-th AGV is active during the k-th event. Using these constraints and the respective notation will lead to the availability time of a seat at the loading zone for the $k + 1$-th event:

$$\bar{v}(k+1) = \max(\bar{x}_1(k+1) + c_1(k+1) + v_1(k+1), \bar{x}_2(k+1) + c_2(k+1) \tag{8.4}$$
$$+ v_2(k+1), \bar{v}(k) + d_8).$$

This model allows the introduction of the following performance constraint for the two redundant AGVs:

$$\bar{v}(k) \leq \bar{v}_{ref}(k). \tag{8.5}$$

This performance constraint allows an identification of the expected availability time for a seat in the loading zone at the k-th event.

This description can then be noted in a condensed form (compare (2.18), (2.19)):

$$\bar{x}(k+1) = A(v(k), v(k+1)) \otimes \bar{x}(k) \oplus B(v(k), v(k+1)) \otimes r(k+1). \tag{8.6}$$

Here $\bar{x} = [\bar{x}_1(k), \bar{x}_2(k), \bar{v}(k)]^T$ and $r(k)$ denote the system input. It is not possible to describe (8.4) directly with max-plus algebra; this problem can be tackled, if two artificial variables are applied. These two virtual variables act as input $r(k)$:

$$r(k) = [\bar{y}_1(k), \bar{y}_2(k)]^T. \tag{8.7}$$

The variable $\bar{y}(k)$ is replaced in (8.1) by either $\bar{y}_1(k)$ or $\bar{y}_2(k)$, which results in:

$$\bar{x}_1(k+1) = \max(\bar{x}_1(k) + b_1(k) + c_1(k), \bar{y}_1(k+1)), \tag{8.8}$$
$$\bar{x}_2(k+1) = \max(\bar{x}_2(k) + b_2(k) + c_2(k), \bar{y}_2(k+1)),$$

with the additional condition:

$$v_i(k+1) = e \Leftrightarrow \bar{y}(k+1) = r_i(k+1) = \bar{y}_i(k+1), \quad i = \{1, 2\}. \tag{8.9}$$

If (8.8) is substituted to (8.4) results in the following form:

$$\bar{v}(k+1) = \max(\bar{x}_1(k) + b_1(k) + c_1(k) + c_1(k+1) + v_1(k+1), \quad (8.10)$$
$$\bar{x}_2(k) + b_2(k) + c_2(k) + c_2(k+1) + v_2(k+1), \bar{v}(k) + d_8,$$
$$\bar{y}_1(k+1) + c_1(k+1) + v_1(k+1),$$
$$\bar{y}_2(k+1) + c_2(k+1) + v_2(k+1)$$

A compact form for the system Eqs. (8.8) and (8.9) can be achieved:

$$A_v = \begin{bmatrix} b_1(k) + c_1(k) & \varepsilon & \varepsilon \\ \varepsilon & b_2(k) + c_2(k) & \varepsilon \\ b_1(k) + c_1(k) + c_1(k+1) + v_1(k+1) & b_2(k) + c_2(k) + c_2(k+1) + v_2(k+1) & d_8 \end{bmatrix}$$
$$(8.11)$$

$$B_v = \begin{bmatrix} e & \varepsilon \\ \varepsilon & e \\ c_1(k) + v_1(k+1) & c_2(k) + v_2(k+1) \end{bmatrix} \quad (8.12)$$

The structure of the these matrices A_v and B_v is depending on the decision variable $v_i(k)$, $v_i(k+1)$; i.e. when the structure of A_v and B_v in the $k+1$-th event is to be obtained, the values of the decision variables $v_i(k)$, $v_i(k+1)$ in k-th and $k+1$-th events need to be taken into account. The processing times ($d_7 = e$, $d_8 = e$) and the transportation times can be introduced, because the matrices A_v and B_v are given, which describe the AGV system. The transportation times can be defined as:

$$\forall k \in \mathbb{N}, i \in \{1, 2\} : c_i(k) = 10, \quad b_i(k) = 10. \quad (8.13)$$

In real manufacturing and assembly systems, frequently the transportation and production times cannot be exactly determined. However, usually an interval of these times can be formulated. For the case $v_1(k+1) = e$ and $v_2(k+1) = \varepsilon$ the structure of matrices A_v and B_v can be formulated with (8.14):

$$A_v = \begin{bmatrix} [18, 22] & \varepsilon & \varepsilon \\ \varepsilon & [18.5, 21.5] & \varepsilon \\ [27.5, 32.5] & \varepsilon & e \end{bmatrix}, \quad B_v = \begin{bmatrix} e & \varepsilon \\ \varepsilon & e \\ [8, 10] & \varepsilon \end{bmatrix}. \quad (8.14)$$

8.4 Modelling of the Seating Assembly System

As stated above, it has several advantages to describe the seating assembly system as "Discrete Event System" (DES). Suitable synchronization rules for the tasks, such as processing and transporting, are one of the main challenges during the design of such complex assembly systems; this topic is addressed in this section.

8.4.1   *Max-Plus Linear Model and Imax-Plus Framework*

It is possible to distinguish two different modes of task synchronization in the seating assembly system. The first mode of task synchronization describes the following scenario: a processing unit under consideration is able to start its intended operation on a next product item (in the $k + 1$-th cycle) when this product item has been transported to this processing unit and when all processing operations on the preceding product item have been completed (in the k-th cycle). In mathematical form, this mode of synchronization can be expressed for the R_1 unit using the following equation:

$$x_1(k + 1) = \max(x_1(k) + d_1, u_1(k + 1) + t_{in,1}) \tag{8.15}$$

The following scenario is described by the second mode of task synchronization: the shared element R_3, which is used within two process chains, can only carry out its intended operation in one of the process chains in a given cycle k—this mode can be called *mutual exclusion mode*. In mathematical form, this mode of synchronization can be expressed for the unit R_3 using the following equations:

$$\begin{aligned}
x_3(k + 1) &= \max(x_1(k + 1) + d_1 + t_{1,3}, x_4(k) + d_4), \\
x_4(k + 1) &= \max(x_2(k + 1) + d_2 + t_{2,3}, x_3(k + 1) + d_3).
\end{aligned} \tag{8.16}$$

On the basis of these assumptions and task synchronization modes, the seat assembly system can be described with the subsequent model:

$$\begin{aligned}
x_1(k + 1) &= \max(x_1(k) + d_1, \ u_1(k + 1) + t_{in,1}) \\
x_2(k + 1) &= \max(x_2(k) + d_2, \ u_2(k + 1) + t_{in,2}) \\
x_3(k + 1) &= \max(x_1(k + 1) + d_1 + t_{1,3}, \ x_4(k) + d_4) = \max(x_1(k) + 2d_1 + t_{1,3}, \ x_4(k) \\
&\quad + d_4, \ u_1(k + 1) + d_1 + t_{in,1} + t_{1,3}) \\
x_4(k + 1) &= \max(x_2(k + 1) + d_2 + t_{2,3}, \ x_3(k + 1) + d_3) = \max(x_1(k) + 2d_1 + d_3 \\
&\quad + t_{1,3}, \ x_2(k) + 2d_2 + t_{2,3}, \ x_4(k) + d_4 + d_3, \ u_1(k + 1) + d_1 + d_3 + t_{in,1} \\
&\quad + t_{1,3}, \ u_2(k + 1) + d_2 + t_{in,2} + t_{2,3}) \tag{8.17} \\
x_5(k + 1) &= \max(x_3(k + 1) + d_3 + t_{3,5}, \ x_5(k) + d_5) = \max(x_1(k) + 2d_1 + d_3 + t_{1,5}, \\
&\quad x_4(k) + d_4 + d_3 + t_{3,5}, \ x_5(k) + d_5, \ u_1(k + 1) + d_1 + d_3 + t_{in1} + t_{1,5}) \\
x_6(k + 1) &= \max(x_4(k + 1) + d_4 + t_{3,6}, \ x_6(k) + d_6 = \max(x_1(k) + 2d_1 + d_3 + d_4 \\
&\quad + t_{1,6}, \ x_2(k) + 2d_2 + d_4 + t_{2,6}, \ x_4(k) + 2d_4 + d_3 + t_{3,6}, \ x_6(k) + d_6, \\
&\quad u_1(k + 1) + d_1 + d_3 + d_4 + t_{in,1} + t_{1,6}, \ u_2(k + 1) + d_2 + d_4 + t_{in,2} + t_{2,6})
\end{aligned}$$

with: $t_{1,5} = t_{1,3} + t_{3,5}, \quad t_{1,6} = t_{1,3} + t_{3,6}, \quad t_{2,6} = t_{2,3} + t_{3,6}.$

It is possible to formulate the preceding equations using max-plus algebra (2.18), (2.19)—this leads to the subsequent system matrices A and B:

$$
A = \begin{bmatrix}
d_1 & \varepsilon & \varepsilon & \varepsilon & \varepsilon & \varepsilon \\
\varepsilon & d_2 & \varepsilon & \varepsilon & \varepsilon & \varepsilon \\
d_1 + t_{1,3} & \varepsilon & \varepsilon & d_4 & \varepsilon & \varepsilon \\
2d_1 + d_3 + t_{1,3} & 2d_2 + t_{2,3} & \varepsilon & d_4 + d_3 & \varepsilon & \varepsilon \\
2d_1 + d_3 + t_{1,5} & \varepsilon & \varepsilon & d_4 + d_3 + t_{3,5} & d_5 & \varepsilon \\
2d_1 + d_3 + d_4 + t_{1,6} & 2d_2 + d_4 + t_{2,6} & \varepsilon & 2d_4 + d_3 + t_{3,6} & \varepsilon & d_6
\end{bmatrix}, \quad (8.18)
$$

$$
B = \begin{bmatrix}
t_{in,1} & \varepsilon \\
\varepsilon & t_{in,2} \\
d_1 + t_{in,1} + t_{1,3} & \varepsilon \\
d_1 + d_3 + t_{in,1} + t_{1,3} & d_2 + t_{in,2} + t_{2,3} \\
d_1 + d_3 + t_{in1} + t_{1,5} & \varepsilon \\
d_1 + d_3 + d_4 + t_{in,1} + t_{1,6} & d_2 + d_4 + t_{in,2} + t_{2,6}
\end{bmatrix}. \quad (8.19)
$$

On the basis of an analytical description of the seat assembly system the subsequent processing and transportation times can be introduced: $d_1 = 1, d_2 = 2, d_3 = 2, d_4 = 2, d_5 = 2, t_5 = 1, t_{in,1} = 2, t_{in,2} = 1, t_{1,3} = 4, t_{2,3} = 1, t_{3,5} = 2, t_{3,6} = 2$.

As mentioned above, an exact determination of the processing and transportation times is usually impossible in industrial practice. It is usually possible to determine interval values of these times, an assumption of them can lead to the subsequent matrices A and B:

$$
A = \begin{bmatrix}
[0, 1] & \varepsilon & \varepsilon & \varepsilon & \varepsilon & \varepsilon \\
\varepsilon & [1.5, 2.5] & \varepsilon & \varepsilon & \varepsilon & \varepsilon \\
[4, 6] & \varepsilon & \varepsilon & [1.5, 2.5] & \varepsilon & \varepsilon \\
[7, 9] & [4, 6] & \varepsilon & [3, 5] & \varepsilon & \varepsilon \\
[9, 11] & \varepsilon & \varepsilon & [5, 7] & [1.5, 2.5] & \varepsilon \\
[11, 13] & [8, 10] & \varepsilon & [7, 9] & \varepsilon & [0.5, 1.5]
\end{bmatrix},
$$

$$
B = \begin{bmatrix}
[1.5, 2.5] & \varepsilon \\
\varepsilon & [0.5, 1.5] \\
[6, 8] & \varepsilon \\
[8, 10] & [3, 5] \\
[10, 12] & \varepsilon \\
[12, 14] & [7, 9]
\end{bmatrix}.
$$

8.4.2 Handling Process Constraints

The precedent deliberations have resulted in a model description. The next step towards a full functionality is the determination of a set of constraints which restrict the behaviour of the seat assembly system. The constraints can be described as follows:

- The system needs to follow a trajectory, which is predefined and may be defined as scheduling constraints of the subsequent form:

$$x_j(k) \leq t_{ref,j}(k), \quad j = 1, \ldots, n-1 \tag{8.20}$$
$$x_n(k) \leq \bar{y}_{ref,i}(k), \quad i = 1, \ldots, n_v. \tag{8.21}$$

In this equation,

$t_{ref,j}(k)$—denotes the upper bound of $t_j(k)$ at the event counter k,
$\bar{y}_{ref,i}(k)$—denotes the upper bound of $y_j(k)$ at the event counter k,
n—denotes the number of the processing (assembly) operation,
n_v—denotes the number of AGVs.

- The first constraint is linked to the second mode of synchronization (mutual exclusion) and enables the framework to avoid possible task conflicts (compare Sect. 8.4.1):

$$\text{if } \exists p_{i,k}, p_{j,k} \in \mathbb{P} \quad \text{then} \quad x_i(k) + d_i - x_j(k) \leq 0 \tag{8.22}$$

where n denotes the system size, i.e. the number of processing units.

- The third constraint is linked to the actual performance of the respective AGV:

$$u_i(k+1) - u_i(k) \leq \bar{u}_i, \quad i = 1, \ldots, r. \tag{8.23}$$

The upper bound $\bar{u}_i$ denotes the maximum speed of the AGV. The energy consumption of the drive motors of the AGV will increase drastically, if this limit is crossed.

- The fourth constraint relates to the change rate:

$$u_j(k+1) - u_j(k) \geq z_j, \quad j = 1, \ldots, r \tag{8.24}$$

where $z_j > 0$ denotes the upper bound of the change rate.

- The final constraint describes the fact that the time for reaching an individual assembly station for $k+1$ cannot be smaller than the time for k.

8.5 Constrained Model Predictive Control

Current automated processes in industry need to fulfil enormous requirements; this leads to difficult process constraints and quality control measures. "Model Predictive Control" (MPC) disposes of the distinct quality that it is able to deal with constraints—consequently the proposed framework is derived from the general MPC strategy for max-plus linear systems as proposed in [2]. The problem boils down to finding the input sequence $\bar{y}(k), \ldots, \bar{y}(k + N_p - 1)$ which will minimise the cost function $J(u)$

$$J(\bar{y}) = -\sum_{j=0}^{N_p-1} \bar{y}(k + j). \tag{8.25}$$

In this equation, N_p denotes the prediction horizon.

It is essential to avoid a direct influence of $x(k + 1), \ldots, x(k + N_p - 1)$ on the scheduling constraints (8.22). This influence can be avoided, if:

$$\tilde{x}(k + N_p - 1) = M \otimes x(k) \oplus H \otimes \tilde{\bar{y}}(k), \tag{8.26}$$

where

$$\tilde{\bar{y}}(k) = \begin{bmatrix} \bar{y}(k + 1) \\ \bar{y}(k + 2) \\ \vdots \\ \bar{y}(k + N_p - 1) \end{bmatrix}, \quad \tilde{x}(k + N_p - 1) = \begin{bmatrix} x(k + 1) \\ \vdots \\ x(k + N_p - 1) \end{bmatrix}.$$

By applying the DES representation (2.18), (2.19) it can be shown that:

$$H = \begin{bmatrix} B_v & \varepsilon & \cdots & \varepsilon \\ A_v \otimes B_v & B_v & \cdots & \varepsilon \\ \vdots & \vdots & \ddots & \vdots \\ A_v^{\otimes N_p-2} \otimes B_v & A_v^{\otimes N_p-3} \otimes B_v & \cdots & B_v \end{bmatrix}, \quad M = \begin{bmatrix} A_v \\ A_v^{\otimes 2} \\ \vdots \\ A_v^{\otimes N_p-1} \end{bmatrix}.$$

It is possible to describe the optimization strategy with the subsequent form: Beginning with an initial condition $x(k)$, the optimal input sequence $\tilde{y}(k)^*$ can be acquired by solving:

$$\tilde{y}(k)^* = \arg \min_{\tilde{y}(k), v_i(k), v_i(k+1)} J(\bar{y}), \tag{8.27}$$

taking into account the constraints (8.2), (8.3) and (8.5).

The precedent elaborations lead to a control algorithm for AGVs with the subsequent structure (compare [5]):

Algorithm 1: max-plus MPC

- Step 0: Set $k = 0$.
- Step 1: Measure the state $x(k)$ and obtain $\tilde{y}(k)^*$ by solving the constrained optimization problem (8.27).
- Step 2: Use the first vector element of $\tilde{y}(k)^*$ (i.e., $\bar{y}(k)^*$) and feed it into the system (2.18), (2.19).
- Step 3: Set $k = k + 1$ and go to *Step 1*.

The optimization problem (8.27) can be characterised as mixed integer-linear programming—current computational solvers can efficiently approach such tasks.

8.6 Fault-Tolerant Control of the Seat Assembly System

In this section, tools are described which can deal with AGV faults (unpermitted deviations of characteristic properties from the nominal condition—compare [1]) which may appear in the seat assembly system. The proposed framework allows designing control strategies for uncertain systems. On the basis of a appropriate system description and general fault scenarios a suitable cost function can be derived which is capable of dealing with complex fault conditions (mechanical and infrastructure issues). The core problem is to identify the input sequence $u(k), \ldots, u(k + N_p - 1)$ which minimises the cost function $J(u)$:

$$J(u) = -\sum_{j=0}^{N_p-1} \sum_{i=1}^{r} q_i u_i(k + j). \tag{8.28}$$

In this equation, $q_i > 0, i = 1, \ldots, m$ is a positive weighting constant which denotes the relative importance of the energy consumption of the i-th AGV and N_p represents the prediction horizon.

However, in some cases the faulty behaviour will lead to an infeasible optimization problem (8.28). In order to deal with these cases, the scheduling constraints should be relaxed:

$$x_j(k) \leq t_{ref,j}(k) + \alpha_j, \quad j = 1, \ldots, n, \tag{8.29}$$

where $\alpha_j \geq 0, j = 1, \ldots, n$ should be rather small in order to lead to a small divergence from the desired time schedule.

In order to find the optimal values of α_j, the subsequent cost function can be used:

$$J(\alpha) = \sum_{i=1}^{n} \alpha_i, \tag{8.30}$$

and the respective optimization framework can be formulated in the following manner:

$$J(u, \alpha) = (1 - \beta) J(u) + \beta J(\alpha). \tag{8.31}$$

In this equation, $0 \leq \beta \leq 1$ is a constant value determined by the development engineer, which can be adjusted to express the higher importance of either $J(u)$ or $J(\alpha)$.

The optimization strategy can be formulated in the following manner

Given an initial condition $x(k)$, find the optimal input sequence $\tilde{u}(k)^*$ by solving:

$$\tilde{u}(k)^* = \arg \min_{\tilde{u}(k), \alpha} J(u, \alpha), \tag{8.32}$$

for the faulty system under constraints (8.22), (8.24) and (8.29). The first step is to avoid a direct influence of $x(k + 1), \ldots, x(k + N_p - 1)$ on the scheduling constraints (8.20). For this purpose, let:

$$\tilde{x}(k + N_p - 1) = M \otimes x(k) \oplus H \otimes \tilde{u}(k), \tag{8.33}$$

where

$$\tilde{u}(k) = \begin{bmatrix} u(k+1) \\ u(k+2) \\ \vdots \\ u(k+N_p-1) \end{bmatrix}, \qquad \tilde{x}(k+N_p-1) = \begin{bmatrix} x(k+1) \\ \vdots \\ x(k+N_p-1) \end{bmatrix} \tag{8.34}$$

Using the description of the DES (2.18), (2.19), it may be shown that:

$$H = \begin{bmatrix} B & \varepsilon & \cdots & \varepsilon \\ A \otimes B & B & \cdots & \varepsilon \\ \vdots & \vdots & \ddots & \vdots \\ A^{\otimes N_p-2} \otimes B & A^{\otimes N_p-3} \otimes B & \cdots & B \end{bmatrix}, \qquad M = \begin{bmatrix} A \\ A^{\otimes 2} \\ \vdots \\ A^{\otimes N_p-1} \end{bmatrix} \tag{8.35}$$

By means of substituting (8.33) into the scheduling constraints (8.29), a linear optimization problem in the form (8.32) can be formulated.

An overview of a predictive FTC scheme system has already been given in Chap. 6. It is possible to explain the developed approach with the underlying FTC algorithm (compare [5]):

Algorithm 2: FTC algorithm

- Step 0: Initialization:
  Set $k = 0$.
- Step 1: Measurement:
  Measure the state $x(k)$ and the sets of the actual production and transportation times: $\mathbb{D}, \mathbb{T}$.

Fig. 8.4 Predictive FTC formalism based on interval max-plus algebra

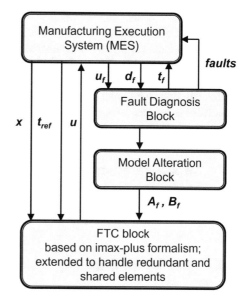

- Step 2: Mobile robot fault diagnosis:
  If $s_i > \delta_i$, then the i-th mobile robot is faulty, where the residual is:

$$s_i = u_f(i, k) - u(i, k)^* \tag{8.36}$$

for all $i = 1, \ldots, r$ and $\delta_i > 0$ being a small positive constant that is robot-depended and should be set by the designer.
- Step 3: Production fault diagnosis:
  On the basis of a set of measurements $\mathbb{D}$ calculate A_f and B_f. If $A_f \geq A$ or $B_f \geq B$ then there is a production fault.
- Step 4: Model alternation
  If there is a production fault then replace A and B with their faulty counterparts A_f and B_f. Additionally, if i-th robot is faulty then replace B by B_f with

$$b_{f,j,i} = b_{j,i} \otimes s_i, \quad j = 1, \ldots, n. \tag{8.37}$$

- Step 5: Solve the linear programming problem under constraints (8.22)–(8.24) and (8.29).
- Step 6: Use the first vector element of $\tilde{u}(k)^*$ (i.e., $u(k)^*$) and feed it into the system.
- Step 7: Set $k = k + 1$ and go to *Step 1*.

In a typical industrial environment, the FTC system can be realised in close connection with the "Manufacturing Execution System" (MES); this is illustrated in Fig. 8.4.

The MES sends the current state of the production tasks $x(k)$ to the FTC block and provides scheduling constraints t_{ref} for the entire prediction horizon N_p. The MES also sends the actual time at which the AGVs reach individual assembly stations (denoted by $u_f(k)$) to the Fault Diagnosis Block. On this basis, the Fault Diagnosis Block can detect a potentially abnormal behaviour of the system and triggers the Model Alternation block to replace the system matrices A and B with their faulty counterparts A_f and B_f. This novel FTC strategy and its interval state space model, which has been explained in Sect. 8.4.1, could be validated on the automated processes in the seat assembly system.

8.7   Implementation Results

This section reports the results of the validation in the seat assembly system.

8.7.1   Fault-Free Case

In this subsection, the performance of the proposed framework is evaluated in the case of a nominal system behaviour. Here, the initial conditions were chosen to be:

$$x(0) = [0, 0, 12.5, 15, 17, 19]^T, \quad \bar{x}(0) = [0, 0]^T. \tag{8.38}$$

The prediction horizon was chosen to be $N_p = 3$. In this case, the scheduling constraints are:

$$
\begin{aligned}
t_{ref}(0) &= [0, 0, 0, 12, 15, 17]^T \\
t_{ref}(1) &= [5, 5, 17.5, 20, 22]^T \\
t_{ref}(2) &= [10, 10, 22.5, 25, 27]^T \\
t_{ref}(3) &= [15, 15, 27, 30, 32]^T
\end{aligned}
\tag{8.39}
$$

$$\cdots$$

and

$$\bar{v}_{ref} = [19, 24, 29, \ldots]. \tag{8.40}$$

The algorithm disposes of a sequential structure and its first step is to determine a sequence of $\bar{y}_{ref}$ which satisfies the scheduling constraint of the AGVs (8.40). Figures 8.5 and 8.6 show the outcome of this initial step.

On the basis of a sequence of y_{ref} that was predicted over N_p the second step of the algorithm can be realised. The control strategy that may allow to follow the reference trajectory is illustrated in Fig. 8.7.

Fig. 8.5 Activity of both AGVs

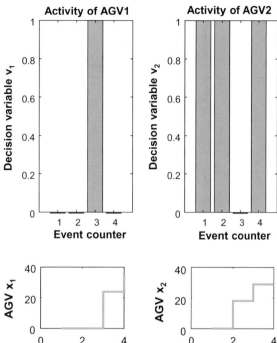

Fig. 8.6 AGV system states and the resulting y_{ref}

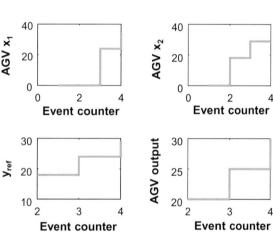

Shared elements are addressed by imposing (8.22). Figure 8.8 illustrates this for resource R_3 (the industrial robot capable of assembling the safety systems).

Figure 8.9 illustrates the seat assembly system output y.

8.7.2 Faulty Case

In this subsection, the reliability of the developed framework is evaluated in the case that faults are present. Here, the subsequent fault scenarios were used in the investigation:

Case 1:   An unpermitted processing time delay of d_6 by 0.75 (e.g. a problem in the assembly station C (for Complex complete seat), which starts from the $k = 3$ event counter.

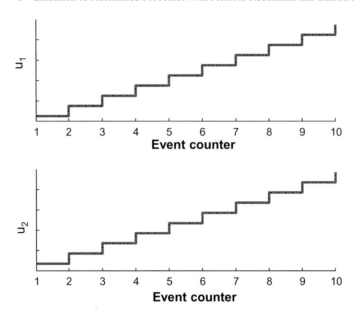

Fig. 8.7 Control strategy for the fault-free case

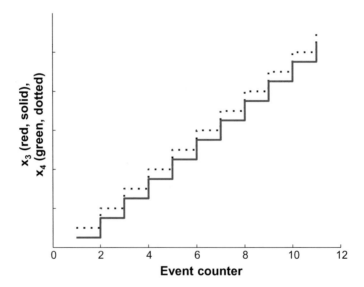

Fig. 8.8 States x_3 and x_4 for the fault-free case

Case 2:   A simultaneous fault of *Case 1* and a second AGV fault (variable u_2) expressed by its delay by 7.5, which starts from the $k = 5$ event counter (e.g. a velocity reduction of the AGV caused by drive motor overheating).

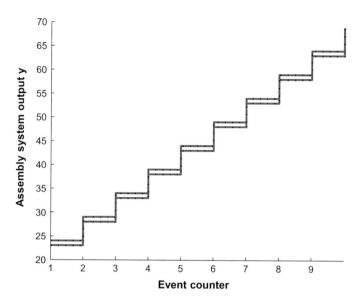

Fig. 8.9 Assembly system output y for the fault-free case

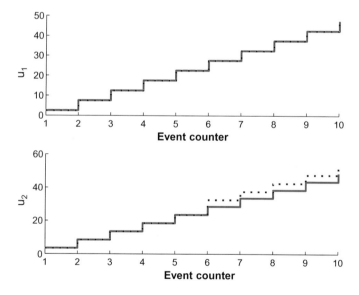

Fig. 8.10 Control strategies with (green dashed) and without FTC (red, solid)

The difference between the FTC and the regular MPC algorithm is illustrated in Fig. 8.10. Figure 8.11 shows the resulting system output y of the seat assembly system.

It is important to note that FTC is able to follow the output y_{ref} resulting in the fault-free case; i.e. the AGVs exhibit analogous behaviour as the one illustrated in

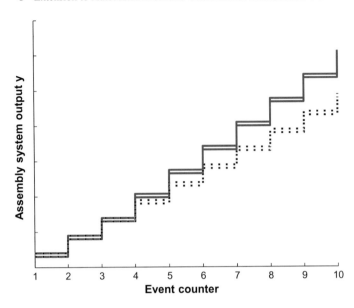

Fig. 8.11 Assembly process output with (green, dashed) and without FTC (red, solid)

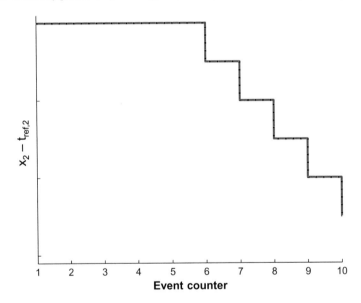

Fig. 8.12 Difference between x_2 and its required trajectory for MPC (red, solid) and FTC (green, dashed)

Figs. 8.5 and 8.6. This behaviour can be achieved by means of realising activities of the assembly station for complex seat frames (resource R_2) before the fault-free schedule, which is presented in Fig. 8.12. It can be seen that x_2 is identical for MPC

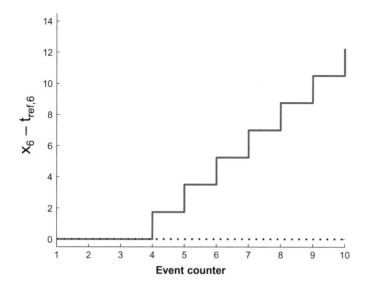

Fig. 8.13 Difference between x_6 and its required trajectory for MPC (red, solid) and FTC (green, dashed)

and FTC but that the control strategy u_2 is contrasting; this is portrayed in Fig. 8.10. Consequently, the FTC can compensate the delays which influence d_6 as shown in Fig. 8.13. A constant performance of the complete seating assembly system can be achieved, because of this delay compensation—even in the case that delays in certain elements are present.

8.8 Conclusions

In the focus of this chapter are flexible redundant and shared elements in automated processes. Due to several specific advantages, both elements are frequently applied in current industrial processes. Flexible redundant elements can increase the reliability and fault-tolerance; shared elements can increase the economic efficiency of automated processes. However, flexible redundant elements require elaborate control systems for task synchronization and shared elements require elaborate control systems for conflict avoidance. It has been demonstrated in this chapter, how interval max-plus fault tolerant control can address these two challenges. An extended framework has been proposed and validated on the example of a seat assembly systems. The results demonstrate also an aptitude for systems with uncertainty as well as a capability to accommodate faults (e.g. reduced velocity causing delays) of system elements such as AGVs and assembly stations. Another appealing fact is that the underlying FTC problem can be solved with the approaches of mixed-linear-integer programming. Future research work will address the application in even more complex systems with a multitude of system elements and fault possibilities.

References

1. Blanke, M., Kinnaert, M., Lunze, J., Staroswiecki, M.: Diagnosis and Fault-Tolerant Control. Springer, New York (2016)
2. de Schutter, T., van den Boom, T.: Model predictive control for max-plus-linear discrete event systems. Automatica **37**(7), 1049–1056 (2001)
3. Komenda, J., Lahaye, S., Boimond, J.-L., van den Boom, T.: Max-plus algebra in the history of discrete event systems. Annu. Rev. Control **45**, 240–249 (2018)
4. Loreto, M., Gaubert, S., Katz, R.D., Loiseau, J.J.: Duality between invariant spaces for max-plus linear discrete event systems. SIAM J. Control. Optim. **48**(8), 5605–5628 (2010)
5. Majdzik, P., Akielaszek-Witczak, A., Seybold, L., Stetter, R., Mrugalska, B.: A fault-tolerant approach to the control of a battery assembly system. Control Eng. Pract. **55**, 139–148 (2016)
6. Majdzik, P., Stetter, R.: A receding-horizon approach to state estimation of the battery assembly system. In: Mitkowski, W., Kacprzyk, J., Oprzedkiewicz, K., Skruch, P. (eds.) Trends in Advanced Intelligent Control Optimization and Automation, pp. 281–290. Springer, Berlin (2017)
7. Manjeet, S., Judd, R.P.: Efficient calculation of the makespan for job-shop systems without recirculation using max-plus algebra. Int. J. Prod. Res. **52**(19), 5880–5894 (2010)
8. Witczak, M., Majdzik, P., Stetter, R., Bocewicz, G.: Interval max-plus fault-tolerant control under resource conflicts and redundancies: application to the seat assembly. Submitt. Int. J. Control (2019)
9. Polak, M., Majdzik, Z., Banaszak, P., Wojcik, R.: The performance evaluation tool for automated prototyping of concurrent cyclic processes. Fundam. Inform. **60**, 269–289 (2004)
10. Seybold, L., Witczak, M., Majdzik, P., Stetter, R.: Towards robust predictive fault-tolerant control for a battery assembly system. Int. J. Appl. Math. Comput. Sci. **25**(4), 849–862 (2015)
11. Tebani, K., Amari, S., Kara, R.: Control of petri nets subject to strict temporal constraints using max-plus algebras. Int. J. Syst. Sci. **49**(6), 1332–1344 (2018)
12. Ting, J., Yongmei, G., Guochun, X., Wonham, W.M.: Exploiting symmetry of state tree structures for discrete-event systems with parallel components. Int. J. Control **90**(8), 1639–1651 (2017)
13. van den Boom, T.J.J., De Schutter, B.: Modelling and control of discrete event systems using switching max-plus-linear systems. Control Eng. Pract. **14**, 1199–1211 (2006)
14. Zhang, R., Cai, K.: Supervisor localisation for large-scale discrete-event systems under partial observation. Int. J. Control **1–13**, (2018)

Chapter 9
Conclusions and Future Research Directions

9.1 Conclusions

An increased fault-tolerance of technical systems is generally demanded by industry and society. This book intends to point out that several elements have to interact in order to achieve the optimum level of fault-tolerance. Firstly, the established algorithms, strategies, methods and tools of "Fault-Tolerant Control" FTC can be applied. It is very sensible to add prognostic approaches to the established steps of FTC in order to be able to include estimations of the remaining useful life into the planning and control of technical processes. It is furthermore sensible to extend the consideration of fault-tolerance into the early stages of a product life—especially to the customer needs exploration and design. This kind of "Fault-Tolerant Design" (FTD) can enhance the controllability of technical systems but can also lead to an incorporation of inherent fault-tolerant characteristics such as fault-tolerance by virtual redundancy.

The algorithms, methods, strategies and tools of FTC and FTD have been explained with the example of automated vehicles and processes. Several causes drive the expansion of both in all areas of human life including industrial production, logistics and infrastructure. Obviously, automated vehicles and processes can lead to economic advantages. Most notably, automated vehicles and processes can also lead to increased safety, reliability and availability, especially if the aspects of fault-tolerance are appropriately considered. Some activities such as the assembly of potentially dangerous goods such as batteries with a high energy density (as used as example here) even require automated vehicles and processes in order to avoid possible consequences of human mistakes. They are also required, if human beings would be exposed to environments with high emissions (e.g. noise) or to the potential occurrence of toxic or otherwise dangerous substances.

The integration of fault-tolerance considerations into the product development can start with methodical processes. One major challenge in such processes is the prioritization of the essential tasks and product modules; several tools have been

© Springer Nature Switzerland AG 2020
R. Stetter, *Fault-Tolerant Design and Control of Automated Vehicles and Processes*, Studies in Systems, Decision and Control 201,
https://doi.org/10.1007/978-3-030-12846-3_9

proposed in this book, which address this challenge, e.g. the project sequence portfolio. One key issue in the product development processes is the management of requirements, i.e. their collection, documentation, structuring, prioritization and tracking. This management can be greatly improved by means of model based techniques. One of the most promising approaches is based on graph-based languages using UML. The main advantage is that such models can be machine executed and allow a thorough connection of requirements into a complete and concurrent product data model.

Automated vehicles and processes need information about their status and their environment in order to operate effectively, efficiently and safely; sensors can deliver this information. Ideally, these sensors can also deliver the information about the presence and absence of faults as well as about their type, location and size—in this case we can refer to them as diagnostic sensors. Innovative approaches can allow to synthesise sensor information in cases where no sensor data are available or the available sensor data are not reliable enough. In such approaches, mathematical models allow to generate reliable sensor information, thus creating virtual sensors. One innovative approach was based on a quadratic boundedness approach and was implemented on a prototype AGV. The application of the approach to AGVs in a fault-free and a faulty driving scenario has been successful. The application made obvious that the resulting estimates exhibit a surprising consistency with the longitudinal forces resulting from a proven reference model. In the case of faults, the resulting estimates immediately generate residuals.

Complex automated processes cannot be totally fault-free, but should still be operating in an optimum manner, even if single or multiple faults occur. Predictive fault-tolerant control systems can present a promising solution for this problem. Novel frameworks, which are based on an interval analysis approach along with max-plus algebra, can allow describing such processes in the most advisable manner as uncertain discrete event systems. Based on this mathematical system description, model predictive control based fault-tolerant strategies can be realised. Such strategies can cope with different kinds of faults concerning processing and transportation and can enable the toleration of the influence of faults on the overall system performance.

The AGVs in automated processes in modern industry are usually battery driven. The useful capacity of these batteries depends both on their state of charge and their state of health. Additionally, the application of AGVs is very often characterised by cooperative work; i.e. more than one AGV can perform a desired task. An innovative framework allows the control of automated processes with awareness of the health of a battery and with cooperating AGVs. The potential of the application has been shown in a seat assembly scenario.

The predictive fault-tolerant control framework can be extended to systems with both *flexible redundant* and *shared* elements; in this case the framework can allow a smooth synchronization of the flexible redundant elements and can avoid resource conflicts for shared elements.

An overview of these main results is given in Fig. 9.1.

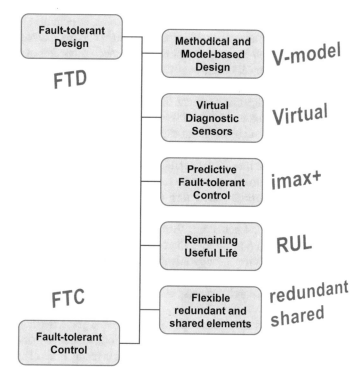

Fig. 9.1 Overview of conclusions

9.2 Future Research Directions

The fault-tolerant design and control of automated vehicles and processes offers multiple possibilities for further research. One prominent topic for future investigations can be the expansion of the strategies, methods and tools for fault-tolerant design and their integration in industrial product development processes. This book has clarified the important aspects, concepts and notions and provided a structure for an in-depth scientific discussion of this relatively new field. Future research projects can add to this fundament and can explore different kinds of product development processes and different kinds of technical systems. Additionally, the existing research concerning "Design for Monitoring" (DfM [1]), "Design for Control" (DfC [3]) and "Design for Diagnosis" (DfD [2]) can be expanded.

During the last decades several research initiatives had been aiming at assisting the process planning and controlling of product development processes of technical systems. This book has structured the different outcomes, has added the specific aspects of fault-tolerant design and control and has emphasised on innovative graph- and model based techniques. More research would be sensible which aims at even more far-reaching and intensive digital modelling approaches for all aspects of technical systems, including aspects regarding physical relationships, structural relationships and logistic considerations as well as monitoring, control and diagnosis tasks. The product development of fault-tolerant complex interconnected systems also requires further attention from science. Additionally, further research can deal with agile forms of process planning and controlling of product development processes and the prominent issue of knowledge management. The optimum placement of sensors and actors for increased fault-tolerance also will offer manifold possibilities for future investigations. Guidelines and tools for an early evaluation of the fault-tolerance of technical systems are another promising field. Researchers together with industry should also explore how the transition of guidelines, tools, algorithms, methods and strategies concerning fault-tolerance can be intensified.

The existing research results concerning virtual diagnostic sensors, which has been addressed in this book, can be accompanied by research initiatives which aim at the development of virtual actuators. Virtual actuators consist of a network of actual actuators and any kind of intelligent compensation unit which can allow the network to fulfil certain functions cooperatively even if one or more actual actuators are faulty. The positive results concerning virtual sensors, which have been described in this book, indicate that research projects concerning virtual actuators could be very fruitful, as well.

Future research activities which can expand the presented predictive health-aware control framework should take even more operating options such as recharging situations into consideration. In this context, further scientific investigations of approaches to address resource conflicts will receive a high merit. Future research should also aim at the fault-tolerant control of manufacturing and assembly systems with less sensors and communication, i.e. systems with unmeasurable and therefore unavailable parameters and times which govern the system. Scientific challenges can still be found in the area of robust fault-tolerant uncertain systems. Consequently, the objectives of future research concerning predictive control need to include robust observers using the presented interval max-plus algebra framework for manufacturing and assembly systems. Future investigations should also include a strengthened analysis of performance-deteriorated parameter data and life-cycle data. A focus of upcoming research can be shared resources, their conscious application and predictive self-diagnosis capabilities for shared resources. Finally, the combination of the predictive control framework with modern technologies such as virtual and augmented reality could lead to improved operator information and optimised planning processes.

Although, future research, as lined out, is desirable, it can still be pointed out that already now the synergistic use of the strategies, frameworks, methods and algorithms described in this book can optimise processes in industry and may lead to technical systems with increased fault-tolerance.

References

1. Stetter, R.: Monitoring in product development. In: Conference Proceedings of the 14th European Workshop on Advanced Control and Diagnosis (ACD) (2017)
2. Stetter, R., Phleps, U.: Design for diagnosis. In: Proceedings of the 18th International Conference on Engineering Design (ICED 11), vol. 5, pp. 91–102 (2011)
3. Stetter, R., Simundsson, A.: Design for control. In: Proceedings of the 21st International Conference on Engineering Design (ICED 17): Design Methods and Tools, vol. 4, pp. 149–158 (2017)

Index

© Springer Nature Switzerland AG 2020
R. Stetter, *Fault-Tolerant Design and Control of Automated Vehicles
and Processes*, Studies in Systems, Decision and Control 201,
https://doi.org/10.1007/978-3-030-12846-3

Printed in the United States
By Bookmasters